For EU product safety concerns:
IN TECH d.o.o., Prolaz Marije Krucifikse Kozulić 3, 51000 Rijeka, Croatia,
info@intechopen.com or visit our website at intechopen.com.

Meet the editor

Dr. Elmer P. Dadios obtained his Ph.D. from Loughborough University, United Kingdom. He is a member of the National Academy of Science and Technology, Philippines. He is a distinguished professor and a university fellow at De La Salle University, Manila, Philippines. He is a professorial chairholder of the Thomas J. Lee Chair in Manufacturing Engineering Management and Victor T. Lu Chair in Production Management. He is the founder and chairman of the board of the Intelligent Systems Innovation Corporation. Dr. Dadios received the 2023 NAST Academician Award, the 2018 Philippine Association for the Advancement of Science and Technology D. M. Consunju Award for Engineering Research, and a Lifetime Achievement Award from the National Research Council of the Philippines (NRCP). He was also recognized among the 50 Men and Women of Science and Technology as well as among Scholar Achievers by the Department of Science and Technology (DOST). Currently, Dr. Dadios is the editor of the *Journal of Advanced Computational Intelligence and Intelligent Informatics* and the *Journal of AI, Computer Science and Robotics Technology*. He is also the editor-in-chief of the *Journal of Computational Intelligence in Engineering Applications.*

Contents

Preface

The concept of fuzzy logic systems came to the fore in 1965 when Prof. Lofti Zadeh introduced the idea of Fuzzy Sets, which were defined as a "class of objects with a continuum of grades of membership." In 1973, Prof. Zadeh also introduced the use of linguistic variables and fuzzy algorithms, a technique that provides an approximate but effective way of describing system characteristics that are too complex and difficult to define and analyze using mathematical models. Prof. Zadeh emphasized that this technique played a very important role in the applications of animate rather than inanimate system constituents' behaviors. In those years, the elites in the academic and scientific community did not initially appreciate the concept of linguistic variables. This is largely because the use of words in systems and decision analysis conflicts with the deep-seated tradition presented by Lord Kelvin in 1883 that gave high regard and respect for numbers in controls. Some critics of Prof. Zadeh's theory include Prof. Rudolf Kalman and Prof. William Kahan, a man with a brilliant mind and an esteemed colleague of Prof. Zadeh. These critics of Prof. Zadeh were proven to be wrong in 1975 when E. H. Mamdami successfully presented an experiment on a fuzzy logic controller for a steam engine of an industrial plant. In 1985, the application of a fuzzy logic controller was solidified when Takagi and Sugeno (TS) published their work, "Fuzzy identification of systems and its applications to modeling and control." The TS method uses linear functions of input variables as the consequence.

The early commercial applications of fuzzy logic were successfully implemented in Japan. In 1987, the Sendai Namboku line began using fuzzy logic to control the subway train's motion, including acceleration and stop operation. The Matsushita vacuums use a fuzzy logic algorithm to adjust suction power automatically. The Hitachi washing machines use fuzzy logic controllers based on weight load, a mixture of fabric, and dirt sensors to automatically set the wash cycle for efficient operation. The Canon camera uses fuzzy logic for its autofocus operation to measure the clarity of the image. The Mitsubishi air conditioners use fuzzy logic to control the room temperature more efficiently than conventional units. Giant corporations like General Motors, Chrysler, Allen Bradley, Whirlpool, Eaton, and Boeing use fuzzy logic systems for efficient use of automotive transmissions, low-power refrigerators, and electric motors. The research on fuzzy logic has grown and flourished, particularly for developing applications in intelligent machines and hybrid controllers. To date, numerous concepts and theories have been formulated for in-depth understanding and advancement of fuzzy logic systems.

This book introduces basic and advanced ideas concerning fuzzy logic systems. It is divided into two sections. The first section (Chapters 1–4) deals with the foundations of the theories and concepts of the fuzzy logic system in advancing technology. The second section (Chapters 5–9) deals with the applications and implementations of these advanced technologies for the benefit of humanity.

Chapter 1, "Review of Type-1 and Type-2 Fuzzy Numbers", emphasizes that fuzzy number theory can be reduced to an argument for interval analysis. It proposes a way of perceiving the concept of fuzzy numbers by comparing it with round numbers.

Chapter 2, "Decoupling of Attributes and Aggregation for Fuzzy Number Ranking", discusses how intuition has been used as a guiding principle for fuzzy number ranking. The chapter adopts the multi-attribute decision-making framework to analyze such intuition.

Chapter 3, "Computing the Performance Parameters of the Markovian Queueing System FM/FM/1 In Transient State", the L–R method to calculate the parameters of performance of the fuzzy Markovian queueing system. The calculation used is the arithmetic of L–R fuzzy numbers restricted to secant approximations. The membership function helps represent graphically the curves of fuzzy parameters' performance in the three-dimensional space of a transient regime in a fuzzy environment.

Chapter 4, "Development of *L*-Group Theory", shows a systematic and successful development of L-group theory. It provides a universal construction of a generated L-subgroup by using level subsets of given L-subsets. This construction allows for defining and studying commutator L-subgroups, normalizer of an L-subgroup, nilpotent L-subgroups, solvable L-subgroups, and normal closure of an L-subgroup. The chapter examines all these concepts and their inter-relationships.

Chapter 5, "Fuzzy Photogrammetric Algorithm for City Built Environment Capturing into Urban Augmented Reality Model", describes and uses Fuzzy Cognitive Maps (FCMs) as a computing framework for matching visual features in an augmented urban-built environment modelling process.

Chapter 6, "PID-like Fuzzy Controller Design for Anti-Slip System in Quarter-Car Robot", proposes a new methodology to control the slip of a Quarter-Car robot using an internal loop based on fuzzy logic inference to compute the gains of a Proportional Integral (PI) structure. The slip is calculated, such as the difference between the linear velocity given by an S-curve velocity profile, and the longitudinal speed is calculated according to the rotational speed of the Quarter-Car tire.

Chapter 7, "Methodology for the Implementation of a Fuzzy Controller on Arduino, MATLAB™ and Nexys 4™ Platforms", discusses a methodology to implement a fuzzy controller in different hardware platforms that can be used to control a process and as an approximator to identify non-linearities and unknown uncertainties of a system.

Chapter 8, "Performance Improvement for Fighter Aircraft using Fuzzy Switching LQI Controller", discusses the switching controller designed for the stabilisation of high-performance aircraft, the Aero-Data Model in Research Environment (ADMIRE). The developed fuzzy logic switching controller has been tested and obtained a robust stabilisation control structure compared to a single conventional LQI and the switched LQI controller.

Chapter 9, "Adaptive Neuro-Fuzzy Inference System-Based GPS-IMU Data Correction for Capacitive Resistivity Underground Imaging with Towed Vehicle System", examines how the Capacitive Resistivity (CR) method utilizes GPS to create maps quickly and with less equipment and labor compared to traditional surveying. However, data acquisition errors can still occur due to GPS sensor accuracy, digital map quality, and map-matching slipups. Also, environmental factors sometimes cause GPS sensors to fail. Hence, reducing errors in GPS receiver accuracy is crucial for correct underground utility location and map matching. This chapter uses an Adaptive Neuro-Fuzzy Inference System (ANFIS) to correct the latitude and longitude positions of a towed vehicle for underground imaging.

Elmer Dadios
Department of Manufacturing Engineering and Management,
De La Salle University,
Manila, Philippines

Section 1

Foundations of the Theories and Concepts of the Fuzzy Logic Systems

Chapter 1

Review of Type-1 and Type-2 Fuzzy Numbers

Norihiro Someyama

Abstract

We review type-1 and type-2 fuzzy numbers in this chapter, and propose one way of perceiving the concept of fuzzy numbers by comparing with that of round numbers. There are some definitions of fuzzy numbers, but we particularly adopt the definition often used in fuzzy analysis. Thereby, we emphasize that fuzzy number theory can be reduced to an argument for interval analysis. Moreover, we explain type-2 fuzzy sets and list two specific type-2 fuzzy numbers, one is a (triangular) perfect quasi-type-2 fuzzy number and the other is a triangular shaped type-2 fuzzy number. Finally, we mention the importance and utility of using type-2 fuzzy numbers.

Keywords: type-*n* fuzzy number, perfect quasi-type-2 fuzzy number, triangular shaped type-2 fuzzy number, membership grade, level-cut set, round number

1. Introduction

A fuzzy number is a special fuzzy set, and the fact that its membership function is unimodal (although "rest stops" along the way up and down the mountain are allowed) and normal is particularly characteristic. The concept of fuzzy sets was introduced by L.A. Zadeh [1] in 1965. There are some definitions of it. We see that in Section 2.3. In each application of fuzzy numbers, you can decide which definition is most useful and choose it on a case-by-case basis. For example, [2] is a well-known reference that defines fuzzy numbers with an eye toward fuzzy analysis (See also [3–6]).

1.1 Where should the concept of fuzzy numbers be used?

The concept of fuzziness or fuzzy numbers appears when we try to distinguish between two or more things or measure something by sight or feeling. If their specific values were required, there would be no need to force fuzziness into the discussion.

For example, coefficients appearing in differential equations can be treated as fuzzy numbers. Let us consider the radiocarbon dating method. Equation describes how several samples of radioactive material decay

$$\frac{\mathrm{d}N}{\mathrm{d}t} = -\lambda N, \quad \lambda > 0, \tag{1}$$

was presented by E. Rutherford. Here $N = N(t)$ is the number of atoms in a radioactive material at time t. Eq. (1) implies that the larger λ, the faster the sample decays. λ varies with the substance, of course, and is determined *experimentally* by the observer on a case-by-case basis (see, e.g., [7] for more information). This λ is set with some mathematical basis, but in some cases it may be determined empirically, or the accuracy of the observation equipment and the skill of the observer may be highly dependent on it. Therefore, it seems more realistic, appropriate and effective to put fuzziness in λ and treat it as a fuzzy number. Differential equations involving fuzzy numbers are generally called fuzzy differential equations. To find the fuzzy solution of a fuzzy differential equation, we can obtain the level-cut sets of the solution by considering its level-cut sets and solving the interval equation, so we can collect them over all levels. However, in order to do so, of course, the fuzzy differential equation must have a (fuzzy) solution. See, e.g., Refs. [3, 6] for the existence and uniqueness of solutions of fuzzy differential equations. Refs. [8–11] are also helpful in knowing how to solve specific fuzzy differential equations of the type as in Eq. (1). In particular, Ref. [8] covers elementary contents of fuzzy numbers and fuzzy differential equations.

1.2 The aim of this chapter

We treat not only fuzzy sets / numbers, but also "fuzzy-membership-grade fuzzy sets / numbers" in this chapter. These are usually called type-2 fuzzy sets / numbers. In order to distinguish fuzzy sets / numbers from these, they are called type-1 fuzzy sets / numbers with emphasis. These are defined and explained in Section 3. We can generally consider type-n fuzzy sets / numbers, but since type-2 fuzzy sets / numbers are used in practical applications, this chapter also deals exclusively with type-1 and type-2 fuzzy sets/numbers.

More precisely, a type-2 fuzzy set can be said to be a fuzzy set whose membership grades are (type-1) fuzzy numbers. For example, it is a fuzzy set such that the membership grade at which a room feels hot is "about 0.8." We can thus think that type-2 fuzzy theory is the application of type-1 fuzzy number theory.

Roughly speaking, the membership function of a type-2 fuzzy set is generally in the shape of a mountain standing above the base of a mountain type. This is because the base is the membership function with "width." The "width" represents the fuzziness of a membership grade (see **Figure 1**). The red curve is the membership function with zero fuzziness.

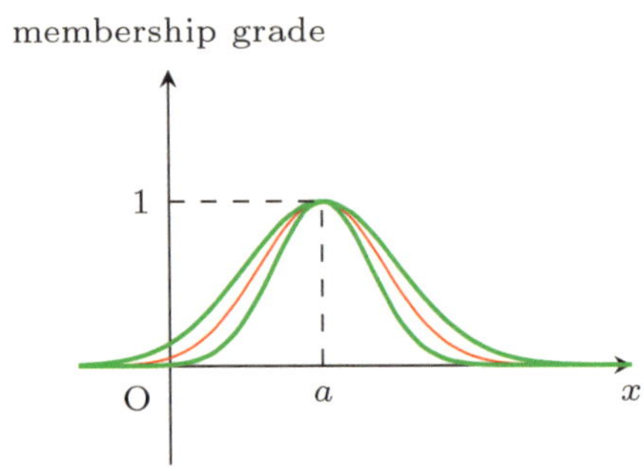

Figure 1.
The form of the "bottom" of a type-2 fuzzy set.

Type-2 fuzzy theory is applied mainly to represent linguistic variables in reasoning. In the first place, Zadeh [1] introduced the concept of type-2 fuzzy sets for this purpose in 1975. By introducing it, we can treat truth values such as "approximately true," "neither true nor false," "truth unknown," etc. This has greatly advanced the study of reasoning (see, e.g., [12–15]). In recent years, type-2 fuzzy number theory has been developing and is being studied in principle as well as in application. Applications to differential equation theory have also been made (see, e.g., [10, 11, 13, 16]. Type-1 fuzzy differential equation theory can be seen in Refs. [3, 6], etc. All of the above studies benefit from the concept of type-1 fuzzy numbers. Indeed, type-2 fuzzy theory is ultimately attributed to type-1 fuzzy theory because any type-2 fuzzy set is represented as a "coupling" of two type-1 fuzzy sets (See Section 3 for details), and hence, concrete computations are also done by level-cutting of type-1 fuzzy sets (see, e.g., [17] for operations for type-2 fuzzy sets). Moreover, the utility of the type-2 fuzzy concept will be explained in Section 3.5.

Therefore, it can be said to be important to interpret the concept of fuzzy numbers appropriately and consider fuzzy numbers whose level-cut sets are easily computed.

From the above, we see the following in this chapter:

- In Section 2, we review how fuzzy numbers are perceived, comparing them to round numbers. Furthermore, we give a "strict" and "suitable" definition of a (type-1) fuzzy number, which reduces its theory to interval number theory and interval analysis [18].

- In Section 3, we meet the concept and some definitions of type-2 fuzzy sets, and know, *via* some figures, that type-2 fuzzy sets / numbers are defined by type-1 fuzzy numbers. Type-2 fuzzy sets /numbers are defined by type-1 fuzzy numbers. Furthermore, we give an example of a type-2 fuzzy number whose level-cut sets are easily computed in Section 3.4 and find out why the type-2 fuzzy concept is necessary or what it can do.

2. Fuzzy numbers, their concept, and level-cut sets

We begin by seeing how to perceive fuzziness on numbers, that is, the concept of fuzzy numbers in this section.

2.1 Difference between round numbers and fuzzy numbers

A fuzzy number, e.g., $\tilde{3}$, is often interpreted and called as "about 3." However, with this representation, it becomes indistinguishable from round numbers, and there is a risk of confusion. We thus verify the difference between round numbers and fuzzy numbers.

Let us consider the following string of positive and finite length (See **Figure 2**). And, let us say that we want to know (even roughly) this length. Then, there are two cases:

a. one is measuring, and

b. the other is eyeballing.

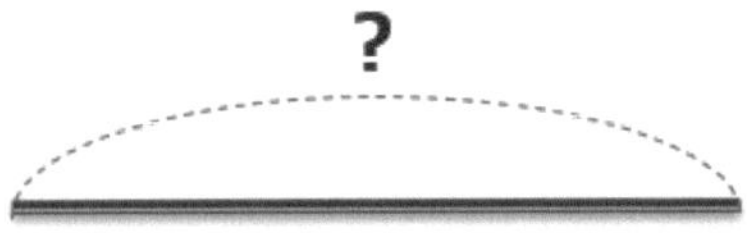

Figure 2.
How long is this string?

Suppose we now have a ruler, tape measure, or other object that can measure length. Then, we can measure the length of the string, and suppose the length is 30.2 cm. This is regarded as about 30 cm, and hence this is case a above.

On the other hand, suppose we do not have a ruler, tape measure, or other object that can measure length. Then, we cannot measure the length of the string, so we need to have an approximate idea of the length, for example, by eyeballing it. Let us say the length feels like "30 cm." Assuming that this visual measurement is perfectly correct, the value 1 (100 % confidence) is assigned to "30 cm." The confidence level gradually decreases as "30 cm" is shifted up or down. We thus represent the barometer of confidence by the graph of a function. This function is called a membership function. This is case b above. Like this, the value from 0 to 1 assigned to each value of length is called the membership grade, often denoted by α. As just explained, the attitude of this chapter is to believe that *membership grade represents a degree of confidence*.

The concept of round numbers appears in the former case, whereas that of fuzzy numbers appears in the latter case. Indeed, if the length can be measured, there is no need to bring in fuzziness, which complicates the discussion. In other words, the difference between round numbers and fuzzy numbers is whether or not an "exact value" exists. The concept of fuzzy numbers has the advantage that it comes into effect when an "exact value" is not available and can be discussed as if it was values out there. The concrete difference is as follows.

- Round number 3 is like $2\sqrt{2}$, 2.9, π, etc. Or, for example, the round number of 10,023 people is 10,000 people. We often call this "about 10,000 people."

- Fuzzy number 3 is the fuzzy set F whose membership function $\mu_F : \mathbb{R} \to [0, 1]$ such that

  i. $\mu_F(3) = 1$,

  ii. μ_F is monotone increasing (resp. deceasing) on $x \in (-\infty, 3)$ (resp. $x \in [3, +\infty)$),

  iii. for example, $\mu_F(x) = 0$ for all $x \leq 2$ and $x \geq 4$.

It seems natural to us that μ_F is continuous, but μ_F can be continuous or discontinuous. Moreover, since any fuzzy set is given based on our subjectivity, there are any numbers of membership functions for it. We give an example of a μ_F in **Figure 3**. It seems more natural that μ_F is smooth, but μ_F can be a broken line as shown in **Figure 3**. Such a fuzzy number is often called a triangular fuzzy number.

The above view of fuzzy numbers is somewhat imprecise and unsuitable for a detailed mathematical discussion. So, a strict definition of fuzzy numbers is given later.

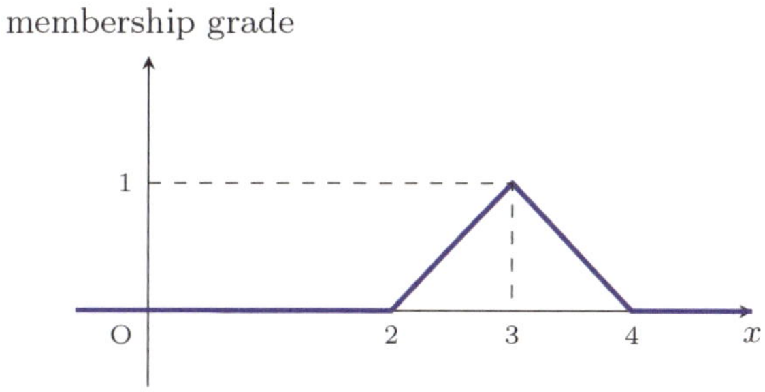

Figure 3.
An example of a membership function of fuzzy number 3.

2.2 Notation of fuzzy sets

In what follows, X denotes the universal crisp set, and we write $A(x) = \mu_A(x)$ for the membership grade of a fuzzy set A at $x \in X$. Related to this, following the conventions in fuzzy analysis, we write $A : X \to [0, 1]$ for a fuzzy set on X.

It is well known that there exist some representations of a fuzzy set A on X:

$$A = \{(x, \mu_A(x)) : x \in X\}, \quad \int_{x \in X} \mu_A(x)/x, \quad \text{etc.} \tag{2}$$

Remark that "$\int$" in the representation on the right side of Eq. (2) means a continuous union for sets, not an integral. Moreover, "/" means a marker, not a division, and "$\int$" is rewritten as "$\sum$" if A is discrete.

2.3 Definitions of fuzzy numbers

There are several definitions for fuzzy numbers. The definition most in line with the senses is as follows.

Definition 2.1. Let $X = \mathbb{R}$ and $A : \mathbb{R} \to [0, 1]$ be a fuzzy set. A is a fuzzy number on $\mathbb{R}$ if and only if A satisfies that

i. A is normal, that is, there exists $x_0 \in \mathbb{R}$ such that $A(x_0) = 1$;

ii. A is convex, that is, $A(tx + (1-t)y) \geq tA(x) + (1-t)A(y)$ for any $x, y \in \mathbb{R}$ and $t \in [0, 1]$;

iii. A is continuous.

An x_0 such that condition i is called a core.

Note: Definition 2.1 says that when considering "about a," the confidence level is 1 (100 %) at $a \in X$ and decreases as the variable $x \in X$ moves away from a to both sides. If $X = \mathbb{Z}$, we should remove condition iii since the membership function is of course discontinuous. In this case, such a fuzzy number is called the discrete fuzzy number.

Recall that any fuzzy set A can be established by its all level-cut sets:

$$A = \bigcup_{\alpha \in [0,1]} \alpha[A]_\alpha, \tag{3}$$

where $[A]_\alpha$ is the α-cut set of A defined as

$$[A]_\alpha = \begin{cases} \{x \in X : A(x) \geq \alpha\} & (\alpha \in (0,1]), \\ \mathrm{supp}(A) = \mathrm{cl}(\{x \in \mathbb{R} : A(x) > 0\}) & (\alpha = 0). \end{cases}$$

Here, "supp" means "support" and cl(S) denotes the closure of a crisp set S. $\alpha[A_\alpha]$ is defined as a fuzzy set *via* the algebraic product operation:

$$(\alpha[A]_\alpha)(x) = \alpha \cdot ([A]_\alpha)(x), \quad x \in \mathbb{R}.$$

Note: There is another way to establish a fuzzy number by its α-cut sets, e.g., Ref. [4]:

$$A = \bigcup_{\alpha \in [0,1]} (\alpha^* \cap [A]_\alpha),$$

where α^* stands for a fuzzy set whose membership function is the constant function, $\alpha^*(x) \equiv \alpha$.

From this, it is expected that the discussion on fuzzy numbers can be reduced to that on intervals (their level-cut sets). But for that we would need a more rigorous definition of fuzzy numbers. We thus adopt the following definition that is often used in fuzzy analysis, etc.

Definition 2.2. Let $u : \mathbb{R} \to [0,1]$ be a fuzzy set. u is a fuzzy number on $\mathbb{R}$ if and only if u satisfies

a. u is normal, that is, u has at least one core;

b. u is fuzzy convex, that is, $u(tx + (1-t)y) \geq u(x) \wedge u(y)$ for any $x, y \in \mathbb{R}$ and $t \in [0,1]$, where $\wedge$ represents the minimum operation;

c. u is semi-upper-continuous, that is, $[u]_\alpha = \{x \in \mathbb{R} : u(x) \geq \alpha\}$ is closed for all $\alpha \in \mathbb{R}$;

d. supp(u) is bounded.

In particular, $u : \mathbb{Z} \to [0,1]$ is called a discrete fuzzy number (on $\mathbb{Z}$).

Definition 2.2 looses Definition 2.1 by replacing conditions ii and iii with conditions b and c, respectively. In fact,

- condition b of Definition 2.2 does not allow the membership function to be bimodal, but allows it to be non-convex (**Figure 4**);

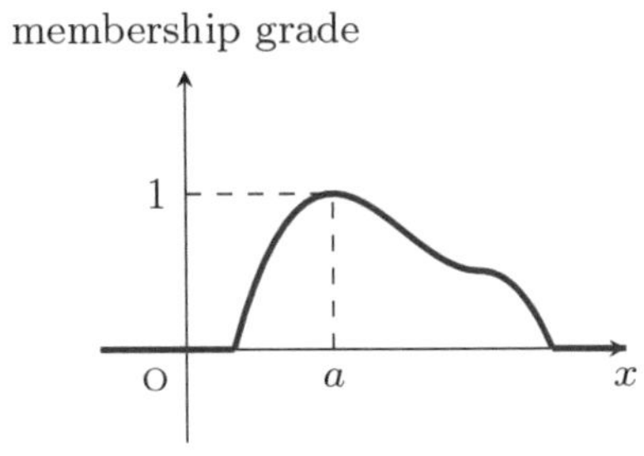

Figure 4.
Disconvexity is OK.

- condition c of Definition 2.2 allows the membership function to be a jump function (**Figure 5**).

On the other hand, however, Definition 2.2 adds a new condition, D, to Definition 2.1. In fact, condition D of Definition 2.2 states that the membership function lands on the x-axis on both sides of core a. Specifically, think of membership functions like the C_0^∞-function

$$u(x) = \begin{cases} \exp\left(-\dfrac{1}{1-x^2}\right) & (|x|<1) \\ 0 & (|x|\geq 1), \end{cases}$$

which is often used as an example of a test function in distribution theory; e.g., (**Figure 6**) [19].

Like the above, there are points in Definition 2.2 where the conditions are loosened or strengthened. The reasons for doing so are discussed in the next subsection.

2.4 Correspondence between fuzzy numbers and level-cut sets

Definition 2.2 implies that u is a fuzzy number if and only if $[u]_\alpha$ is a bounded closed interval for any $\alpha \in [0,1]$. In fact,

- condition a says that $[u]_\alpha$ is not empty for any $\alpha \in [0,1]$,

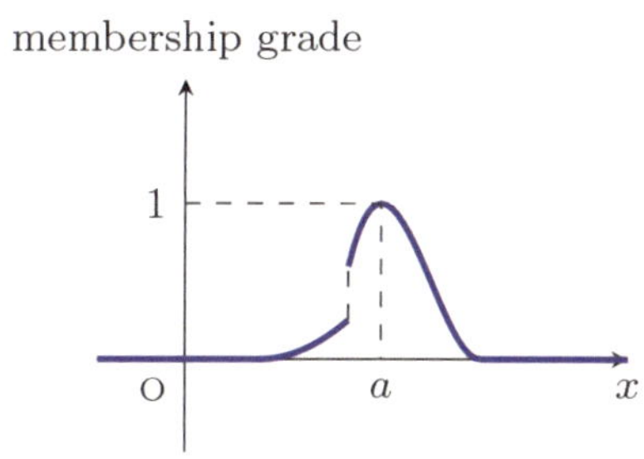

Figure 5.
Jump is OK.

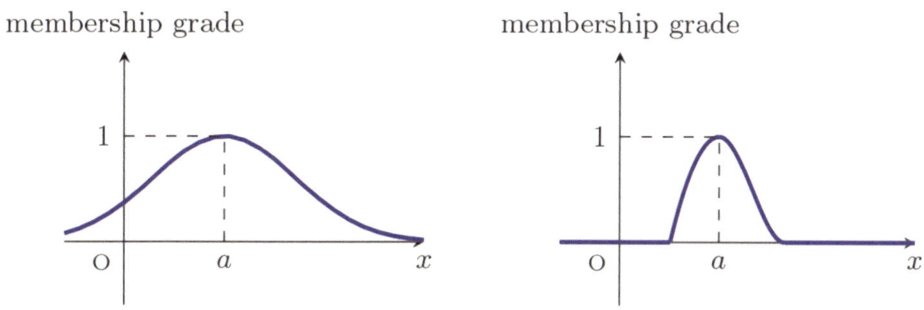

Figure 6.
Definition 2.1 allows membership functions like the left side, but Definition 2.2 requires membership functions like the right side.

- condition b says that $[u]_\alpha$ is an interval on $\mathbb{R}$ for any $\alpha \in [0,1]$,
- condition c says that $[u]_\alpha$ is closed for any $\alpha \in [0,1]$,
- condition d says that $[u]_\alpha$ is bounded for any $\alpha \in [0,1]$.

Given this, we denote

$$[u]_\alpha = [u_-(\alpha), u_+(\alpha)]$$

for any $\alpha \in [0,1]$. $u_\pm(\alpha)$ are, of course, crisp numbers (or, the crisp function with respect to α, $[0,1] \ni \alpha \mapsto u_\pm(\alpha) \in \mathbb{R}$). Hence, we expect that the discussion of fuzzy number theory can be reduced to that of interval analysis, that is, level-cut theory.

Theorem 2.3 (Representation Theorem for Fuzzy Numbers, e.g. [2–4, 6]) Let $u : \mathbb{R} \to [0,1]$ be a fuzzy number. Then, the following holds:

1. $[u]_\alpha$ is a non-empty bounded closed interval for any $\alpha \in [0,1]$.
2. If $0 \le \alpha_1 \le \alpha_2 \le 1$, then $[u]_{\alpha_2} \subset [u]_{\alpha_1}$.
3. For the monotone increasing and positive sequence $\{\alpha_n\}$ that converges to $\alpha \in (0,1]$, one has

$$\bigcap_{n=1}^{\infty} [u]_{\alpha_n} = [u]_\alpha.$$

Conversely, if there is a family of sets $\{P_\alpha\}_{\alpha \in [0,1]}$ satisfying properties 1, 2 and 3 above, then there exists a unique fuzzy number u. Moreover, it follows that

$$[u]_\alpha = P_\alpha$$

for any $\alpha \in (0,1]$ and

$$[u]_0 \subset P_0. \tag{4}$$

Note: As can be seen, Eq. (4) does not guarantee equality. For example, we want to treat fuzzy numbers u in the same way as crisp numbers if possible, so it is necessary to define the four arithmetic operations, etc., for fuzzy numbers. To do so, we only need to well define the operations of level-cut sets (i.e., interval numbers), which corresponds to P_α above. It must then be satisfied that $P_\alpha = [u]_\alpha$ for any $\alpha \in [0,1]$. Representation Theorem and the proof of $P_0 \subset [u]_0$ guarantee that the results of the defined operations are fuzzy numbers. Eq. (3) guarantees that $\cup_{\alpha \in [0,1]} \alpha P_\alpha$ is a fuzzy set, but does not guarantee that it is a fuzzy number. For this reason, what is needed is the representation theorem. This is detailed in, e.g., Ref. [4].

3. Type-2 fuzzy numbers: Two concrete examples

3.1 Motivation for type-2 fuzzy theory

The key to fuzzy theory is the concept of membership grades, and it is represented by our individual degrees of confidence. However, membership grades we set may be ambiguous. Put another way, determining membership grades means determining the degrees of confidence, but there is also the degree of confidence for the degree of confidence α. For example, let us say that we are willing to accept the sentiment that we have determined α (100α %) with a confidence level of 0.8 (80 %). These are called fuzzy membership grades. In summary, type-2 fuzzy theory is a theory of fuzzy sets with fuzzy membership grades.

A type-1 fuzzy set A is characterized by its membership function $A : X \to [0,1]$ for the universal (crisp) set X, whereas a type-2 fuzzy set $\tilde{A}$ is characterized by its membership function $\tilde{A} : X \to [0,1]^{[0,1]}$. Here, U^V denotes the set of mappings $U \to V$ for crisp sets U, V.

There are other advantages of type-2 fuzzy theory. See Section 3.5 for that.

3.2 Type-2 fuzzy sets and those associated with them

We begin with the definition of a type-2 fuzzy set. Type-2 fuzzy theory has many concepts and their terms, but we prepare them required in this chapter. (There is a slight change from the traditional definition and notation.) For example, the membership grade of a type-2 fuzzy set is called the fuzzy membership grade of it. Unlike type-1 fuzzy sets, type-2 fuzzy sets can be three-dimensional figures and are generally difficult to depict.

Definition 3.1. If $\tilde{A}$ is characterized by the membership function

$$\mu_{\tilde{A}} : I \times J_x \ni (x,u) \mapsto \mu_{\tilde{A}}(x,u) \in [0,1], \tag{5}$$

$\tilde{A}$ is called a type-2 fuzzy set on X. Here, $I \subset X$ is the universe for the primary variable $x \in X$, and $J_x \subset [0,1]$ is the interval determined for each $x \in I$. Then, I and J_x are called the primary and secondary domains of $\tilde{A}$, respectively.

There are other representations of $\tilde{A}$:

$$\tilde{A} = \left\{ \left(x, u; \mu_{\tilde{A}}(x,u)\right) : x \in I,\ \ u \in J_x \right\}, \quad \int_{x \in I} \int_{u \in J_x} \mu_{\tilde{A}}(x,u)/(x,u), \quad \text{etc.} \tag{6}$$

Remark that "$\int$" in the representation on the right side of Eq. (6) means a continuous union for sets, not an integral. Moreover, "/" means a marker, not a division, and "$\int$" is rewritten as "$\sum$" if $\tilde{A}$ is discrete.

Definition 3.2. Let $\tilde{A}$ be a type-2 fuzzy set on X. The type-1 fuzzy set for $\tilde{A}$ appears if $x \in I$ is fixed arbitrarily. It is called the vertical slice of $\tilde{A}$, and its membership function

$$\nu_{\tilde{A}}^{x} : J_x \to [0,1].$$

is called the secondary membership function of $\tilde{A}$ at x.

$\tilde{A}$ can be said to be characterized by $\nu_{\tilde{A}}^{x}$ as follows:

$$\tilde{A} = \int_{x \in X} \left(\int_{u \in J_x} \nu_{\tilde{A}}^{x}(u)/u \right) /x,$$

where $\nu_{\tilde{A}}^{x}(u)$ is the value of the secondary membership function at u, that is, the secondary membership grade of $\tilde{A}$.

Note: The concepts of vertical slices and secondary membership functions are often treated in the same sense. Because of this, "secondary membership functions" are sometimes also called "vertical slices."

There are two kinds of cutting with respect to β and what is important is what to cut with respect to β. First, the following is the cutting of vertical slices.

Definition 3.3. Let $\tilde{A}$ be a type-2 fuzzy set on X. The crisp set

$$S_{\tilde{A}}(x|\beta) = \begin{cases} \left\{u \in J_x : \nu_{\tilde{A}}^x(u) \geq \beta\right\} & (\beta \in (0,1]), \\ \mathrm{cl}\left(\left\{u \in J_x : \nu_{\tilde{A}}^x(u) > 0\right\}\right) & (\beta = 0), \end{cases}$$

is called the β-cut set of the vertical slice of $\tilde{A}$ (**Figure 7**).

Next, the following is the cutting of type-2 fuzzy sets.

Definition 3.4. Let $\tilde{A}$ be a type-2 fuzzy set on X. For $\beta \in [0, 1]$,

$$\tilde{A}_\beta = \bigcup_{x \in I} S_{\tilde{A}}(x|\beta)$$

is called the β-plane of $\tilde{A}$. In particular, $\tilde{A}_1$ and $\tilde{A}_0$ are called the principal (or, principle) set and footprint (set) of $\tilde{A}$, respectively.

Roughly speaking, a type-2 fuzzy set can be characterized by a membership function in the form of two mountains. We use the following notation for the β-plane of $\tilde{A}$ because we want to make it geometrically easy to see what a type-2 fuzzy set looks like. That is, it brings in the notion of "left- and right-sided type-1 fuzzy sets."

Definition 3.5. If the β-plane of $\tilde{A}$ is the interval-valued fuzzy set, there exist type-1 fuzzy sets $\underline{A_\beta}$ and $\overline{A_\beta}$. Then, we denote

$$\tilde{A}_\beta = \left\langle \underline{A_\beta},\ \overline{A_\beta} \right\rangle.$$

Here, $\underline{A_\beta}$ and $\overline{A_\beta}$ are called the lower membership function (briefly, LMF) and upper membership function (briefly, UMF) on $\tilde{A}$, respectively.

Definition 3.6. Let $\tilde{A}$ be a type-2 fuzzy set on X. For each $\beta \in [0, 1]$, the coupling of α-cut sets of $\underline{A_\beta}$ and $\overline{A_\beta}$ is written as

$$[\tilde{A}]_\beta^\alpha = \left\langle \left[\underline{A_\beta}\right]_\alpha,\ \left[\overline{A_\beta}\right]_\alpha \right\rangle, \quad \alpha \in [0, 1], \tag{7}$$

and is called the (α, β)-cut set of $\tilde{A}$.

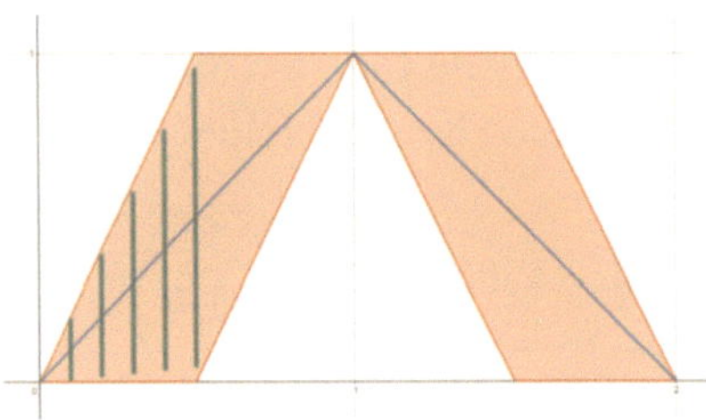

Figure 7.
0-cut set of the vertical slice of "about 1" as the meaning of type-2 fuzzy numbers; ***Figure 1*** *[11].*

As with type-1, we can discuss type-2 fuzzy sets as their β-planes or (α, β)-cut sets. Indeed, Hamrawi found the formula that is a type-2 version of Eq. (3) as follows; we leave the details to Ref. [20] for more information on the contents of this neighborhood.

Proposition 3.7. Any type-2 fuzzy set $\tilde{A}$ on X satisfies

$$\tilde{A} = \bigcup_{\beta \in [0,1]} \beta \bigcup_{\alpha \in [0,1]} \alpha \left[\tilde{A}\right]_{\beta}^{\alpha},$$

where $\alpha \left[\tilde{A}\right]_{\beta}^{\alpha} : X \to \{0, \alpha\}$ is a type-1 fuzzy set.

3.3 Perfect quasi-type-2 fuzzy numbers

We hereafter set $X = \mathbb{R}$.

Hamrawi introduced the following type-2 fuzzy number, which we can call a "triangular type-2 fuzzy number."

Definition 3.8. ([20], Section 3.4). Let $\mathcal{A}$ be a type-2 fuzzy set on $\mathbb{R}$. $\mathcal{A}$ is a perfect type-2 fuzzy number if and only if

i. UMF and LMF of FP($\mathcal{A}$) are equal as type-1 fuzzy numbers, and

ii. UMF and LMF of P($\mathcal{A}$) are equal as type-1 fuzzy numbers.

Moreover, if a perfect type-2 fuzzy number $\mathcal{A}$ satisfies that

iii. $\mathcal{A}$ can be completely determined by using its FP($\mathcal{A}$) and P($\mathcal{A}$),

such a $\mathcal{A}$ is called the perfect quasi-type-2 fuzzy number (briefly, PQT2FN) on $\mathbb{R}$.

Definition 3.9. A PQT2FN $\mathcal{A}$ is triangular if and only if $[\mathcal{A}]_{\beta}^{\alpha}$ has the α-cut set of LMF on $\mathcal{A}$:

$$\begin{aligned}
\left[\underline{A}_{\beta}\right]_{\alpha} &= \left[L_{\underline{A}_{\beta}}^{\alpha},\ R_{\underline{A}_{\beta}}^{\alpha}\right]; \\
L_{\underline{A}_{\beta}}^{\alpha} &= X_{A_1}^{\alpha} - (1-\beta)\left(X_{A_1}^{\alpha} - L_{\underline{A}_0}^{\alpha}\right), \\
R_{\underline{A}_{\beta}}^{\alpha} &= Y_{A_1}^{\alpha} + (1-\beta)\left(R_{\underline{A}_0}^{\alpha} - Y_{A_1}^{\alpha}\right), \\
L_{\underline{A}_0}^{\alpha} &= C_{\mathcal{A}} - (1-\alpha)\left(C_{\mathcal{A}} - L_{\underline{A}_0}\right), \\
R_{\underline{A}_0}^{\alpha} &= C_{\mathcal{A}} + (1-\alpha)\left(R_{\underline{A}_0} - C_{\mathcal{A}}\right)
\end{aligned}$$

and the α-cut set of UMF on $\mathcal{A}$:

$$\begin{aligned}
\left[\overline{A}_{\beta}\right]_{\alpha} &= \left[L_{\overline{A}_{\beta}}^{\alpha},\ R_{\overline{A}_{\beta}}^{\alpha}\right]; \\
L_{\overline{A}_{\beta}}^{\alpha} &= X_{A_1}^{\alpha} - (1-\beta)\left(X_{A_1}^{\alpha} - L_{\overline{A}_0}^{\alpha}\right), \\
R_{\overline{A}_{\beta}}^{\alpha} &= Y_{A_1}^{\alpha} + (1-\beta)\left(R_{\overline{A}_0}^{\alpha} - Y_{A_1}^{\alpha}\right), \\
L_{\overline{A}_0}^{\alpha} &= C_{\mathcal{A}} - (1-\alpha)\left(C_{\mathcal{A}} - L_{\overline{A}_0}\right), \\
R_{\overline{A}_0}^{\alpha} &= C_{\mathcal{A}} + (1-\alpha)\left(R_{\overline{A}_0} - C_{\mathcal{A}}\right),
\end{aligned}$$

where

$$X^{\alpha}_{A_1} = C_{\mathcal{A}} - (1-\alpha)(C_{\mathcal{A}} - X_{A_1}),$$
$$Y^{\alpha}_{A_1} = C_{\mathcal{A}} + (1-\alpha)(Y_{A_1} - C_{\mathcal{A}}).$$

They are called the left principle number and right principle number of $\mathcal{A}$, respectively. $C_{\mathcal{A}}$ denotes the core of $\mathcal{A}$, that is, the crisp number $[\mathcal{A}]^1_1$. A triangular perfect quasi-type-2 fuzzy number is abbreviated as TPQT2FN.

Figure 8 shows

$$L^{\alpha}_{\overline{A}_0} \leq X^{\alpha}_{A_1} \leq L^{\alpha}_{\underline{A}_0} \leq C_{\mathcal{A}} \leq R^{\alpha}_{\underline{A}_0} \leq Y^{\alpha}_{A_1} \leq R^{\alpha}_{\overline{A}_0}.$$

In particular, the supports of $\mathcal{A}$ are represented by the α-cut sets of LMF and UMF of FP($\mathcal{A}$):

$$[\underline{A}_0]_{\alpha} = \left[L^{\alpha}_{\underline{A}_0}, R^{\alpha}_{\underline{A}_0}\right], \quad [\overline{A}_0]_{\alpha} = \left[L^{\alpha}_{\overline{A}_0}, R^{\alpha}_{\overline{A}_0}\right].$$

Also, the α-cut set of P($\mathcal{A}$) is given as

$$[A_1]_{\alpha} = \left[X^{\alpha}_{A_1}, Y^{\alpha}_{A_1}\right].$$

Now, recall that the triangular type-1 fuzzy number u is determined by three information, that is, its left end l, core c and right end r:

$$u = \langle\langle l; c; r\rangle\rangle.$$

In contrast, TPQT2FN $\mathcal{A}$ is determined by seven information, that is, its upper left end $L_{\overline{A}_0}$, left principle number X_{A_1}, lower left end $L_{\underline{A}_0}$, core $C_{\mathcal{A}}$, lower right end $R_{\underline{A}_0}$, right principle number Y_{A_1} and upper right end $R_{\overline{A}_0}$. We then write

$$\mathcal{A} = \left\langle\left\langle L_{\overline{A}_0}, X_{A_1}, L_{\underline{A}_0}; C_{\mathcal{A}}; R_{\underline{A}_0}, Y_{A_1}, R_{\overline{A}_0}\right\rangle\right\rangle.$$

In general, any type-2 fuzzy set/number satisfies that both the principal set and the vertical slice are type-1 fuzzy numbers as shown in **Figure 9**.

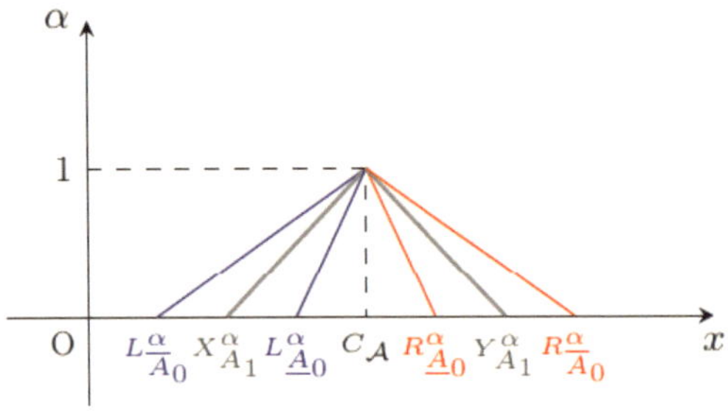

Figure 8.
A view of a TPQT2FN $\mathcal{A}$ from directly above.

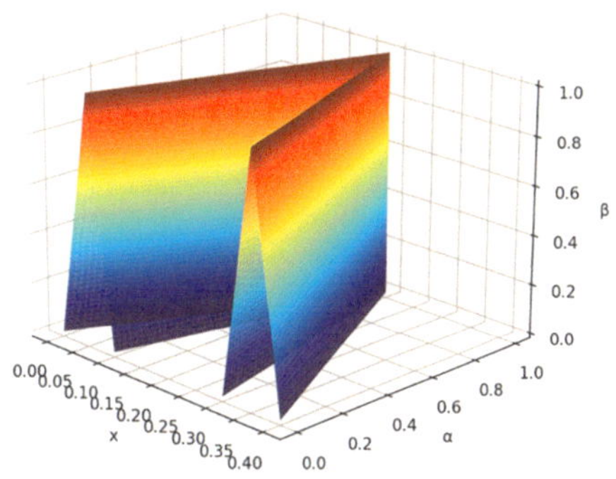

Figure 9.
Membership function of a perfect quasi-type-2 fuzzy number.

3.4 Triangular shaped type-2 fuzzy numbers

The (α, β)-cuts of a perfect quasi-type-2 fuzzy number can be easily obtained, but at cost of its condition being too strict (too ideal). We want to consider a more natural type-2 fuzzy number while still being able to easily compute the α-cuts. H. Uesu proposed the following type-2 fuzzy number, and he and the author, et al. [11] introduced in 2022.

Definition 3.10. Let $\tilde{A}$ be a type-2 fuzzy set whose core is $a \in \mathbb{R}$ on $\mathbb{R}$. $\tilde{A}$ is a triangular shaped type-2 fuzzy number (briefly, TST2FN) on $\mathbb{R}$ if and only if its principal set and secondary membership function at x are given by

$$\tilde{A}_1(x) = \max\{1-|x-a|, \ 0\},$$

$$\nu_{\tilde{A}}^x(t) = \begin{cases} 1 - \dfrac{|t - \tilde{A}_1(x)|}{\min\{\tilde{A}_1(x), 1 - \tilde{A}_1(x)\}} & (\tilde{A}_1(x) \in (0,1)), \\ 1 & (t = \tilde{A}_1(x) \in \{0,1\}), \\ 0 & (t \neq \tilde{A}_1(x) \in \{0,1\}), \end{cases}$$

respectively (**Figure 10**).

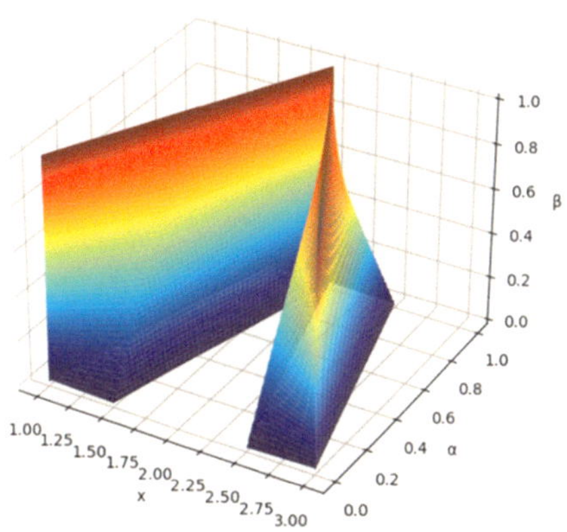

Figure 10.
Membership function of a triangular shaped type-2 fuzzy number with core 2; ***Figure 4*** *[11].*

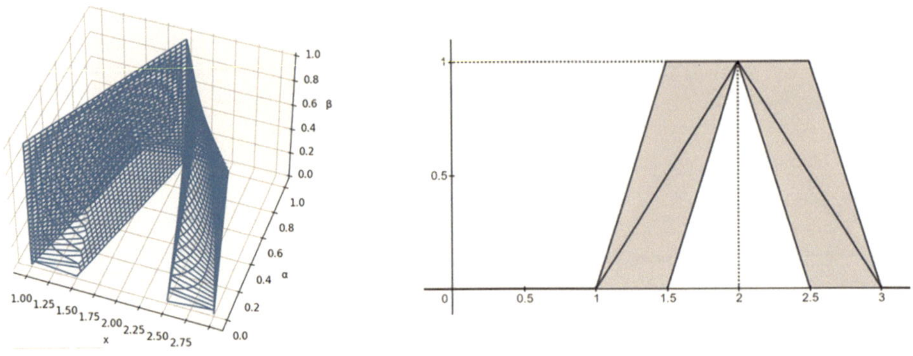

Figure 11.
Footprint set of "about 2" as the meaning of type-2 fuzzy numbers; ***Figure*** *5 [11].*

Note: For any TST2FN $\tilde{A}$, both its left and right footprints are congruent parallelograms of width-length 0.5 (See **Figure 11**). Moreover, the diagonals of the two parallelograms constitute the principal set $\tilde{A}_1$ of $\tilde{A}$ (see the right sides of **Figures 11** and **12**).

Even a fuzzy number by its natural definition is not suitable for application if the computation of its level-cut sets is complex. However, a TST2FN is defined naturally and its level-cut sets can be easily computed.

Theorem 3.11 ([11], Theorem 2.19) Let $\tilde{A}$ be a TST2FN with core $a \in \mathbb{R}$. For $\alpha, \beta \in [0, 1]$, the following holds:

$$[\tilde{A}]_\beta^\alpha = \begin{cases} \left\langle \left[a - \dfrac{2-\alpha-\beta}{2-\beta},\ a - \dfrac{\beta-\alpha}{\beta}\right], \left[a + \dfrac{\beta-\alpha}{\beta},\ a + \dfrac{2-\alpha-\beta}{2-\beta}\right] \right\rangle, & 0 < \alpha \le \dfrac{\beta}{2}; \\ \left\langle \left[a - \dfrac{2-\alpha-\beta}{2-\beta},\ a - \dfrac{1-\alpha}{2-\beta}\right], \left[a + \dfrac{1-\alpha}{2-\beta},\ a + \dfrac{2-\alpha-\beta}{2-\beta}\right] \right\rangle, & \dfrac{\beta}{2} < \alpha \le 1 - \dfrac{\beta}{2}; \\ \left\langle \left[a - \dfrac{1-\alpha}{\beta},\ a - \dfrac{1-\alpha}{2-\beta}\right], \left[a + \dfrac{1-\alpha}{2-\beta},\ a + \dfrac{1-\alpha}{\beta}\right] \right\rangle, & 1 - \dfrac{\beta}{2} < \alpha \le 1. \end{cases} \tag{8}$$

As we can see from the two type-2 fuzzy numbers above, it may be said that a type-2 fuzzy number is determined by

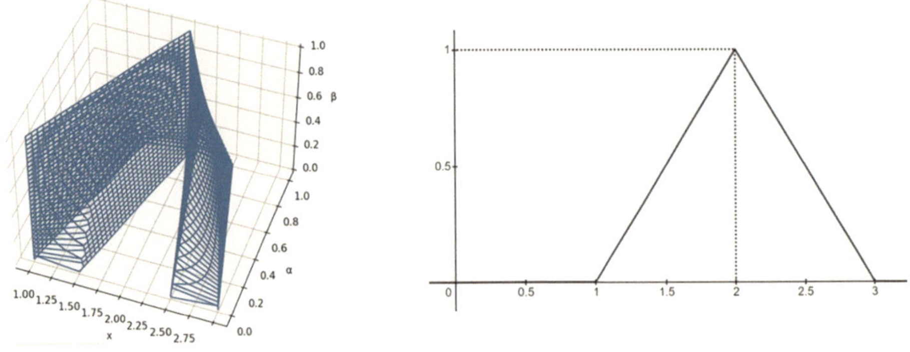

Figure 12.
Principal set of "about 2" as the meaning of type-2 fuzzy numbers; ***Figure*** *6 [11].*

- what form the FP should take and
- how to connect that FP to the PS.

In case of PQT2FNs, the FP is in triangular form, and the FP and the PS are linearly connected. In case of TST2FNs, the FP is in the shape of a parallelogram, and the FP and the PS are curvilinearly connected.

3.5 Utility of the concept of type-2 fuzzy numbers

Type-2 fuzzy theory has an advantage. For example, as discussed in Ref. [9], there are cases where the observer is a veteran or a newcomer to some experiment. In such a case, the coefficients appearing in fuzzy differential equations, etc., may change depending on the former and the latter (in case of Eq. (1), we are talking about the value of λ). Actually, it can be considered that PS and FP of a type-2 fuzzy set correspond, so to speak, the veteran who make no mistakes at all and the newcomer with no experience at all, respectively. Hence, by discussing type-2 fuzzy theory, we can have the discussion of fuzziness concluded the case of an experiment by any observer (Veteran or not!).

Some readers may think that instead of going to the trouble of discussing type-2 fuzzy numbers, they can simply consider two type-1 fuzzy numbers and compare them. However, doing so would result in obtaining two fuzzy numbers under separate environments (conditions), and it would generally not make sense to compare them, for example. In other words, depending on the nature of the research, one may wish to compare multiple subjects under the same conditions as appropriately as possible. The type-1 fuzzy theory of comparing by each membership function does not, however, seem to be appropriate in general. In fact, we often establish membership functions under unique conditions of the subjects, and hence, the subjects are compared under different conditions. The way of this research will not give appropriate comparison results.

Then, type-n fuzzy theory is useful in overcoming this problem. With type-n fuzzy numbers, membership functions or level-cut sets for all objects under the same conditions can be obtained simultaneously (see Eqs. (7) and (8) as the case $n = 2$).

When comparing two objects, the number of times to obtain level-cut sets is the same whether considering two type-1 fuzzy numbers or one type-2 fuzzy number, but basically type-2 fuzzy numbers are more likely to be computationally expensive or unobtainable with respect to the level-cut sets. However, it is not very effective to consider a type-2 fuzzy number that is too convenient for us only because it is easier to calculate. Therefore, when dealing with type-2 fuzzy numbers, we prefer to consider something that is easy to calculate while still being in accordance with our senses. One example of this is TST2FN, Definition 3.10. In addition, one of the applications of this can be seen in Ref. [21].

4. Conclusions

We regarded membership grades as the "degrees of confidence" in this chapter. In particular, Ref. [22] is well known for this same idea of literature.

Although the application aspect is important in fuzzy theory, this chapter focused on how to recognize fuzzy numbers in the first place rather than how to apply them.

This is because if the starting point of the discussion is iffy, so will the outcome. Fuzzy theory is not a theory to derive fuzzy results, but the ability to approach fuzziness mathematically (with some rigor) is the real appeal of fuzzy theory. That is, the beginning is crucial, and this mindfulness encourages the proper application of fuzzy numbers.

This chapter reviewed

- how to perceive and find "fuzziness,"
- that to discuss fuzzy numbers, it is sufficient to discuss their level-cut sets,
- that we should define or introduce fuzzy numbers such that their level-cut sets are easily obtained in order to do that.

These things are common to type-n fuzzy numbers for any $n \in \mathbb{N}$.

We have redefined the concept of fuzzy numbers by comparing them to round numbers. A round number is a number whose exact value is known and whose value is replaced by a tractable number. On the other hand, a fuzzy number is a concept that attempts to estimate its value when the exact value is (forever) unknown and determine its membership function. Even using the same word "about," they are different in concept itself, let alone approach. In summary, fuzzy numbers are a valid concept for quantities that definitely exist but whose values are difficult to obtain.

Furthermore, the discussion of fuzzy numbers can be reduced to that of interval analysis. Instead of dealing directly with fuzzy numbers, we can discuss them by dealing with the level-cut sets, which are (nonempty bounded closed) intervals. Hence, we want to treat fuzzy numbers whose level-cut sets are easily obtained. Compared to type-1 fuzzy numbers, type-2 fuzzy numbers are generally more complex and difficult to find for their level-cut sets, and in particular, type-2 fuzzy numbers we should be dealing with must be easy to compute. With this in mind, TST2FNs were introduced in Ref. [11]. The application of TST2FNs to type-2 fuzzy differential equation theory can be also seen in [11]. If we want to know about type-2 fuzzy differential equation theory, we can also see in, e.g., [9, 16, 23].

The computation of fuzzy numbers tends to be more tedious and complicated than that of crisp numbers. Therefore, it is not sufficient to make anything a fuzzy number if it is an ambiguous number. It is necessary to appropriately determine what should be regarded as a fuzzy number, even at the expense of calculation tediousness and complexity. One criterion for such a judgment was given in this chapter.

Acknowledgements

I would like to particularly thank Dr. Jiro Inaida. I learned the rigorous formulation of arithmetic, etc. for fuzzy numbers by level-cut sets from him. I also deeply thank my co-researcher, senior and friend, Dr. Hiroaki Uesu. My research on type-2 fuzzy numbers is largely due to him.

DOI: http://dx.doi.org/10.5772/intechopen.110495

Author details

Norihiro Someyama
Shin-yo-ji Buddhist Temple, Tokyo, Japan

*Address all correspondence to: philomatics@outlook.jp

References

[1] Zadeh LA. Fuzzy Sets. Information and Control. 1965;**8-3**:338-353

[2] Diamond P, Kloeden P. Metric Spaces of Fuzzy Sets: Theory and Applications. World Scientific; 1994

[3] Gomes LT, de Barros LC, Bede B. Fuzzy Differential Equations in Various Approaches. Springer (Springer Briefs in Mathematics); 2015

[4] Inaida J. Fuzzy Numbers. Tokyo-kougei-sha (Summary note on the serial papers "Fuzzy Numbers I-IV" in Japan Society for Fuzzy Theory and Intelligent Informatics, in Japanese). 2022. p. 28.

[5] Inaida J. Taylor series on the fuzzy number space. Biomedical Soft Computing and Human Sciences. 2009;**16**(1):15-25

[6] Lakshmikantham V, Mohapatra RN. Theory of Fuzzy Differential Equations and Inclusions. Taylor & Francis (Series in Mathematical Analysis and Applications); 2003. p. 6

[7] Burghes DN, Borrie MS. Modelling with Differential Equations (Mathematics and its Applications). Ellis Horwood; 1981

[8] Buckley JJ, Eslami E, Feuring T. Fuzzy Mathematics in Economics and Engineering. Springer (Studies in Fuzziness and Soft Computing); 2002

[9] Mazandarani M, Najariyan M. Differentiability of Type-2 fuzzy number-valued functions. Communications in Nonlinear Science and Numerical Simulation. 2014;**19**: 710-725

[10] Someyama N, Uesu H, Shinkai K, Kanagawa S. Type-2 fuzzy initial value problems for second-order T2FDEs. Preprint, arXiv:2101.10324v2 [math. GM]. 2021:27

[11] Someyama N, Uesu H, Shinkai K, Kanagawa S. Triangular shaped Type-2 fuzzy numbers: Application to Type-2 fuzzy differential equations. Ann. Fuzzy Math. Inform. 2022;**24**(1):29-52

[12] Mendel JM. Advances in Type-2 fuzzy sets and systems. Information Sciences. 2007;**177-1**:84-110

[13] Mendel JM, Liu F. On new quasi-type-2 fuzzy logic systems. In: Proceedings of 2008 IEEE International Conference on Fuzzy Systems (FUZZ 2008), WCCI 2008. Hong Kong, China; 2008. p. 7

[14] Zadeh LA. The concept of a linguistic variable and its application to approximate reasoning - I. Information Sciences. 1975;**8**:199-249

[15] Zhai D, Mendel JM. Uncertainty measures for general Type-2 fuzzy sets. Information Sciences. 2011;**181-3**:503-518

[16] Bandyopadhyay A, Kar S. On Type-2 fuzzy partial differential equations and its applications. Journal of Intelligent & Fuzzy Systems. 2018;**34**:405-422

[17] Karnik NN, Mendel JM. Operations on Type-2 fuzzy sets. Fuzzy Sets and Systems. 2001;**122**:327-348

[18] Moore RE, Kearfott RB, Cloud MJ. Introduction to interval analysis. Society for Industrial and Applied Mathematics Philadelphia. 2009. p. 184

[19] Lieb EH, Loss M. Analysis. American Mathematical Society (Graduate Studies in Mathematics), Vol. 14. Second Edition. 2001.

[20] Hamrawi H. Type-2 Fuzzy Alpha-Cuts. De Montfort University. Ph D thesis; 2011. p. 177

[21] Uesu H. Triangular shaped Type-2 fuzzy number and UESU product. ICIC Express Letters. 2022;**16-8**:869-876

[22] Kaufmann A, Gupta MM. Introduction to Fuzzy Arithmetic. Van Nostrand Reinhold Company; 1991

[23] Bandyopadhyay A, Kar S. System of Type-2 fuzzy differential equations and its applications. Neural Computing and Applications. 2018;**31**:5563-5593

Chapter 2

Decoupling of Attributes and Aggregation for Fuzzy Number Ranking

Simon Li

Abstract

Intuition, expressed as verbal arguments or axiom formulations, has often been used as a guiding principle for fuzzy number ranking (FNR). This chapter adopts the multi-attribute decision making (MADM) framework to analyze such intuition with three results. First, intuition in FNR should have involved multiple attributes, which are often implicated in the existing ranking methods. Then, we suggest three attributes (i.e., representative x-value, x-value range, overall membership ratio), which can be used to characterize the FNR intuition. Second, we decouple two issues in FNR: selection of attributes and aggregation of values, where aggregation is concerned with the trade-off among attributes to determine a single index for FNR. Then, the discount factors are proposed for the attributes of range and membership ratio to model the trade-off and formulate a ranking index. Third, the decoupling of attributes and aggregation reveals a fundamental tension between information content and the satisfaction of the FNR axioms. That is, if we can consider more information (in terms of attributes) as relevant to FNR, the ranking method will likely violate some FNR axioms. However, if we consider less information, the ranking method will be less sensitive to distinguish some fuzzy numbers for ranking. In the end, the proposed multi-attribute approach can provide a practical aspect to analyze and address the FNR problems.

Keywords: fuzzy number ranking, multi-attribute decision making, aggregation, decoupling, overall membership ratio

1. Introduction

Intuition has been a criterion for researchers to evaluate and comment the ranking results from a set of fuzzy numbers. As a pattern described by Wang and Kerre [1], a ranking method can be criticized by yielding "counter-intuitive" results from some examples, and thus it is motivated to develop new ranking methods (e.g., [2, 3]). Despite of its common use, the meaning of "intuition for ranking" is somewhat unclear. It should be related to the ranking of real numbers, which is fundamental in our intuition. However, this alone is not sufficient for fuzzy number ranking (FNR). Why? When we compare two real numbers: 3 and 5, we can state 5 > 3 because these real numbers can be ordered on a single dimension, i.e., the real line. In the context of fuzzy sets, the membership information is added. For example, consider two ordered pairs: (3, 0.9) and (5, 0.4), where the second elements are the membership values. Here, we cannot straightforwardly state (5, 0.4) ≻ (3, 0.9) due to the presence of the second dimension, membership, in the ranking consideration. Notably, in this paper, the symbol ">" is used to compare two real numbers, while the symbol "≻" or "≽" is used to represent the ranking relation.

Consider that a fuzzy number contains a set of such ordered pairs. We argue that the problem structure of FNR should contain multiple dimensions to explain "intuition" properly. Then, we employ the classical framework of multi-attribute decision making (MADM) [4] for the analysis of ranking intuition. The framework of MADM distinguishes the concepts of attributes and aggregation. Attributes are used to evaluate the properties of options, and they are subject to the selection by decision makers, who determine what properties (or information) are deemed relevant to the decision problems. On the other hand, aggregation captures the weighting strategies (e.g., weighted sum) to address the trade-off consideration among the option's properties (or information).

In FNR, each attribute represents a single dimension for ranking consideration. In literature, numerous attributes have been implied in the formulations of ranking indices. For example, the approach of the maximizing and minimizing sets [3, 5, 6] implicates the attributes that articulate the optimistic and pessimistic aspects of a fuzzy number for ranking. In the centroid-based approach [7, 8], centroid can be interpreted as an attribute that focuses on the "middle" aspect over the geometry of a fuzzy number. Notably, each notion of attribute can be quantified in multiple ways. For example, we may express the notion of "average" via the formulations of "value" by Delgado et al. [9] or "median" by Bodjanova [10]. In addition, new ranking methods have been proposed by adding attributes to the ranking indices. For example, to address some non-distinguishable results from Abbasbandy and Hajjari [2], Asady [11] and Ezzati et al. [12] formulated additional attributes (namely, the epsilon-neighborhood and $Mag'(u)$, respectively) in their ranking indices.

The consideration of multiple attributes for FNR is not new. In literature, some approaches have explicitly considered multiple measures (or attributes) to describe a fuzzy number such as value and ambiguity [9, 13, 14], mean and standard deviation [15], average value and degree of deviation [16], expected value (in transfer coefficient) and deviation degree [17, 18], general concepts of area/mode/spreads/weights [19–21] and extensions from the centroid concept [21–25].

Aggregation is a separate issue from the selection of attributes. It aims to handle the given information of attributes for decision making. In literature, different aggregation approaches over the same attributes have been reported. For example, aggregation over the x- and y-coordinates of a centroid can be done via a distance measure [26] or an area measure [8, 27]. The weighted sum approach has been used to aggregate two attributes such as the right/left utility values [3] and the average and deviation values [16]. In addition to closed-form equations, aggregation can also be done by rules and procedures. For example, Asady [11] and Chi and Yu [23] determine the ranking of fuzzy numbers based on the priority of two or three attributes, which basically is a lexicographical ordering procedure ([4], pp. 77–79).

The aggregation approach can influence the ranking results since it controls the trade-off among attributes. To illustrate, consider the earlier ordered pairs (3, 0.9) and (5, 0.4). Suppose that two attributes are considered for ranking: real number and membership value, and we assume "higher value → higher rank" for both attributes. Then, we can have multiple ways to aggregate these two values such as 3 + 0.9 and

3×0.9, which are consistent with the "higher-the-better" direction. However, different aggregation functions can lead to different ranking results, e.g., $(3 + 0.9) < (5 + 0.4)$ and $(3 \times 0.9) > (5 \times 0.4)$. Different results can be explained by the trade-off approach implied in the aggregation functions. For example, in this case, addition tends to give an advantage to real number, whereas multiplication allows more influence from membership value.

Based on the above discussion, the theme of this chapter is to adopt the MADM framework, which purposely decouples attributes and aggregation for FNR. In this way, we can compare ranking methods in view of their selections of attributes and the formulations of aggregation functions independently. In addition, the multi-attribute aspect can help explain the axiomatic properties of ranking methods. To avoid the reliance on the "intuition criterion", Wang and Kerre [1] suggested seven axioms as reasonable properties to specify the meaning of intuition more clearly. Ban and Coroianu [28] derived a class of ranking functions that can satisfy six of these axioms with literature examples that can belong to this class under some conditions (e.g., [2, 29, 30]).

Despite of the formal work by Ban and Coroianu [28], new ranking methods emerge continually as researchers considered this class of ranking functions did not address two aspects. First, the development by Ban and Coroianu [28] was intended for normalized fuzzy numbers, and some work has been developed for the non-normalized cases (e.g., [31]). Second, their class of ranking functions cannot distinguish two symmetric fuzzy numbers with different spreads (e.g., for cases in Ezzati et al., [12]). In some recent work, Dombi and Jónás [32] applied the probability-based preference intensity index, and Van Hop [33] developed the dominant interval measure (namely relative dominant degree) for fuzzy number ranking. Their approaches basically generalized the numerical techniques of intervals for fuzzy number ranking without decomposing or analyzing the ranking attributes.

More fundamentally, it seems to us that if a ranking function is designed to satisfy the axioms by Wang and Kerre [1], this ranking function will be less sensitive to the distribution of membership values and the spreads of fuzzy numbers to determine the ranking results. In other words, the satisfaction of these axioms is strongly influenced by the type of information (or attributes) that is selected for FNR but it is less relevant to the aggregation approach. The distinction between information selection and aggregation has not been investigated for fuzzy number ranking in literature. This chapter will use the multi-attribute aspect to analyze this issue.

After the preliminaries in Section 2, this chapter will discuss and illustrate our selection of three attributes for FNR in Section 3 and then our aggregation approach using the discount factors in Section 4. Section 5 will suggest some guidance for the application of the proposed multi-attribute ranking method. Section 6 will discuss the relation between the information content for FNR and the axiomatic properties of ranking methods. This chapter is concluded in Section 7.

2. Preliminaries

Fuzzy number is described as a fuzzy subset of the real line $\mathbb{R}$ [34]. This work considers trapezoidal fuzzy number (TrFN) as a special case of fuzzy number. Let F_A denote a TrFN with a maximum membership equal to h_A, as illustrated in **Figure 1**. Let x be any element of the real line, and its membership according to TrFN, denoted as $\mu_{F_A}(x)$, can be expressed in the following formulation where a_1, a_2, a_3, a_4 are real

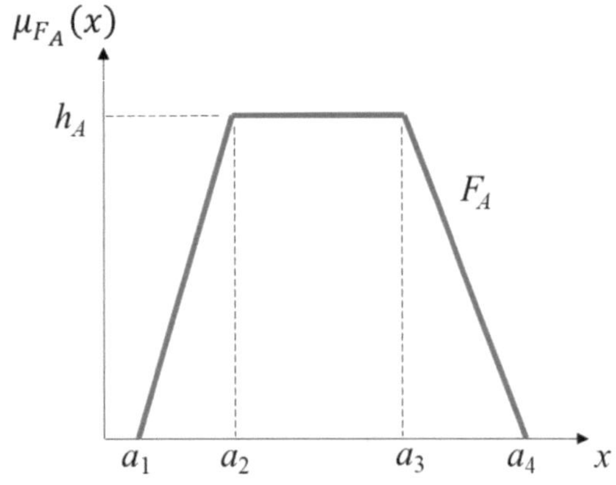

Figure 1.
A trapezoidal fuzzy number.

numbers to specify F_A. As a convenient notation, F_A can be expressed as a 5-tuple, where $F_A = (a_1, a_2, a_3, a_4; h_A)$.

$$\mu_{F_A}(x) = \begin{cases} 0 & x < a_1 \\ \left(\dfrac{x - a_1}{a_2 - a_1}\right) h_A & a_1 \le x < a_2 \\ h_A & a_2 \le x < a_3 \\ \left(\dfrac{x - a_4}{a_3 - a_4}\right) h_A & a_3 \le x < a_4 \\ 0 & x > a_4 \end{cases} \tag{1}$$

Let $supp(F_A)$ be the support of F_A, and we have $supp(F_A) = \{x \in \mathbb{R} \mid a_1 \le x \le a_4\}$. Then we have the infimum and supremum of $supp(F_A)$ as $\inf supp(F_A) = a_1$ and $\sup supp(F_A) = a_4$, respectively. Also, let $I_{F_A}(\alpha) = [l_{F_A}(\alpha), r_{F_A}(\alpha)]$ be the α-cut interval of F_A. For $\alpha \le h_A$, the left and right bounds of the α-cut interval can be formulated as follows.

$$l_{F_A}(\alpha) = a_1 + \left(\frac{\alpha}{h_A}\right)(a_2 - a_1) \tag{2}$$

$$r_{F_A}(\alpha) = a_4 + \left(\frac{\alpha}{h_A}\right)(a_3 - a_4) \tag{3}$$

Suppose we have two fuzzy numbers: $F_A = (a_1, a_2, a_3, a_4, h_A)$ and $F_B = (b_1, b_2, b_3, b_4, h_B)$, and a constant, denoted as λ (i.e., $\lambda \in \mathbb{R}$). We can have fuzzy number addition and multiplication with a constant as follows [17, 18, 34].

$$F_A \oplus F_B = (a_1 + b_1, a_2 + b_2, a_3 + b_3, a_4 + b_4;\ min\,\{h_A, h_B\}) \tag{4}$$

$$\lambda \cdot F_A = (\lambda \cdot a_1, \lambda \cdot a_2, \lambda \cdot a_3, \lambda \cdot a_4; h_A) \tag{5}$$

To describe some reasonable properties of ranking methods, Wang and Kerre [1] have proposed seven axioms. Ban and Coroianu [28] have dropped one axiom by considering a ranking (or an ordering) over a given set of fuzzy numbers. This chapter follows the choice made by Ban and Coroianu [28]. Let $\boldsymbol{F}$ be a set of fuzzy numbers, and a ranking method determines the binary relation $\succcurlyeq$ over $\boldsymbol{F}$. Then, their six axioms are summarized (without elaborating their variants) below.

Axiom 1: $F_A \succcurlyeq F_A$. for any $F_A \in \boldsymbol{F}$.

Axiom 2: For any F_A, $F_B \in \boldsymbol{F}$, if $F_A \succcurlyeq F_B$ and $F_B \succcurlyeq F_A$, then $F_A \sim F_B$.

Axiom 3: For any F_A, F_B, $F_C \in \boldsymbol{F}$, if $F_A \succcurlyeq F_B$ and $F_B \succcurlyeq F_C$, then $F_A \succcurlyeq F_C$.

Axiom 4: For any F_A, $F_B \in \boldsymbol{F}$, if $\inf supp(F_A) \geq \sup supp(F_B)$, then $F_A \succcurlyeq F_B$.

Axiom 5: Suppose that F_A, F_B, $F_A \oplus F_C$, $F_B \oplus F_C$ are elements of $\boldsymbol{F}$. If $F_A \succcurlyeq F_B$, then $F_a \oplus F_c \succcurlyeq F_b \oplus F_c$.

Axiom 6: Suppose that $\lambda \in \mathbb{R}$ and F_A, F_B, $\lambda \cdot F_A$, $\lambda \cdot F_B$ are elements of $\boldsymbol{F}$. If $F_A \succcurlyeq F_B$ and $\lambda \geq 0$, then $\lambda \cdot F_A \succcurlyeq \lambda \cdot F_B$. If $F_A \succcurlyeq F_B$ and $\lambda \leq 0$, then $\lambda \cdot F_A \preccurlyeq \lambda \cdot F_B$.

Axioms 1 and 3 are referred to as the reflexive and transitive properties of binary relations, respectively, for a total pre-order on $\boldsymbol{F}$ [28]. Axiom 2 defines the conditions for the equality "$\sim$". Axiom 4 specifies that F_A is larger than or equal to F_B if the lower bound of the support of F_A is larger than the upper bound of the support of F_B. Axioms 5 and 6 generally imply that the ordering of $F_A \succcurlyeq F_B$ should be preserved if they are added by the same fuzzy number F_C or multiplied by the same positive quantity λ. Notably, index-based ranking methods will satisfy Axioms 1 to 3 [1], and this chapter will focus more on Axioms 4 to 6.

3. Three attributes for fuzzy number ranking

In this section, we characterize the comparison of fuzzy numbers through one primary attribute and two secondary attributes. The primary measure is concerned with the representative value of a fuzzy number on the real line, which is a common intuition for ranking. One secondary attribute checks the range of real numbers enclosed by a fuzzy number, which information is independent of the representative value but can be relevant for ranking. Another secondary attribute is associated with membership, which is concerned with the shape of a fuzzy number.

3.1 Representative x-value

Since the real line of a fuzzy number is often expressed on the x-axis, we use "x-value" to label the values associated with the real line. As a fuzzy number encloses a range of possible x-values, one common intuition is to identify a representative x-value of a fuzzy number for comparison. There can be several options that are aligned with this intuition such as the expected value [34, 35], the x-coordinate of a centroid [36] and median [10]. In this chapter, we adopt the class of ranking indices derived by Ban and Coroianu [28]. Let $rep(F_A, w)$ be the function to evaluate the representative x-value of the fuzzy number F_A, and its formulation is given as follows.

$$rep(F_A, w) = w \cdot a_1 + \left(\frac{1}{2} - w\right)a_2 + \left(\frac{1}{2} - w\right)a_3 + w \cdot a_4 \tag{6}$$

where w is a weighting constant with $0 \leq w \leq 1$. As proven by Ban and Coroianu [28] (Theorem 39), if this function is used as a ranking index, it satisfies the six axioms discussed in the preliminaries section. Beyond this theorem result, we can interpret this formulation as a weighted function of a fuzzy number's core values (a_2 and a_3) with a weight (1/2-w) and boundary values (a_1 and a_4) with a weight w. When w = 0, only the core values are considered. Alternately, when w = 1/2, only the boundary values are considered. To emphasize the importance of core values (i.e., a_2 and a_3) through weighting, we set $(1/2\text{-}w) \geq w$, and then we have $0 \leq w \leq 1/4$.

Derived from Eq. (6), we have $rep(F_A, w) \geq rep(F_B, w)$ if the following condition is satisfied.

$$w[(a_1 + a_4) - (b_1 + b_4)] + \left(\frac{1}{2} - w\right)[(a_2 + a_3) - (b_2 + b_3)] \geq 0 \tag{7}$$

This condition implies a weighted comparison between core and boundary values of F_A and F_B. Apparently, we cannot guarantee the satisfaction of this condition if F_A and F_B are partially overlapped (i.e., $supp(F_A) \cap supp(F_B) \neq \varnothing$). Then, the value of w can influence the ordering of $rep(F_A, w)$ and $rep(F_B, w)$. Following the discussion in Ban and Coroianu [28], we consider the presence of w as a generalization of some existing indices, which have implicitly pre-defined weighting factors for core and boundary values of a fuzzy number. For example, the ranking index developed by Abbasbandy and Hajjari [2] is an instance by setting w = 1/12. Given a ranking problem, decision makers can consider some sensitivity analysis (e.g., evaluate the value of w that makes $rep(F_A, w)$ = $rep(F_B, w)$) to define the value of w for their ranking problems.

3.2 X-value range

Another attribute is associated with the range of possible x-values of a fuzzy number. Fuzzy numbers can have the same representative x-values with different ranges (e.g., symmetric triangular fuzzy numbers with the same core value but different boundary values). Some argue that the information of range should be considered for ranking (e.g., [11]). There can be several options to quantify this intuition such as ambiguity value [9, 13], standard deviation [15] and deviation degree [16–18]. In this chapter, we adopt the range (or size) of the α-cut interval (denoted as $rng(F_A, \alpha)$), and it is formulated as follows.

$$rng(F_A, \alpha) = r_{F_A}(\alpha) - l_{F_A}(\alpha) \tag{8}$$

Figure 2 illustrates the α-cut interval of a trapezoidal fuzzy number F_A., where the lower (left) and upper (right) bounds of the α-cut interval are denoted as $l_{F_A}(\alpha)$ and

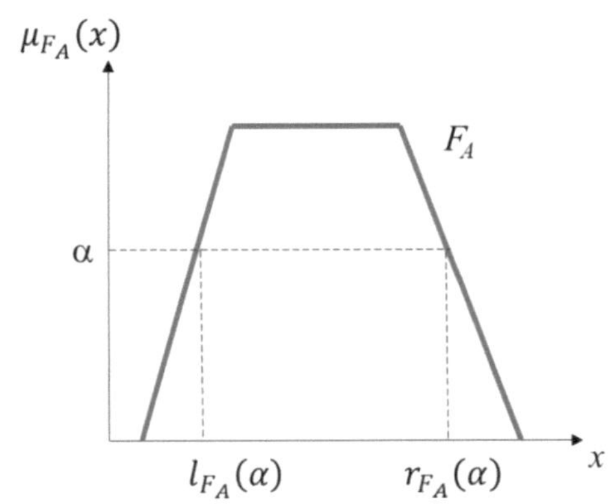

Figure 2.
Illustration of the α-cut interval.

$r_{F_A}(\alpha)$, respectively. The formulations of $l_{F_A}(\alpha)$ and $r_{F_A}(\alpha)$ can be found in Eqs. (2) and (3), respectively. The value of α can be interpreted as the minimum membership value that is deemed relevant for the ranking analysis. For example, if we set α at a lower value, we will receive a wider interval.

Here, we suppose that a large range of possible x-values tends to yield a lower rank because decision makers do not want high uncertainty associated with a large range. This stated intuition of "larger range ➔ lower rank" is aligned with Wang and Luo [6] and Nasseri et al. [37]. Also, we classify range as a secondary attribute because some decision makers may find this attribute not necessary to their ranking problems (e.g., ranking a set of triangular fuzzy numbers with a similar size of support). Then, using the measure of representative x-value only could be sufficient for ranking. In contrast, if decision makers find the information of range relevant to their ranking problems, our suggested approach is to take the range information as a modifier to the representative x-value. This approach will be discussed in Section 5.

3.3 Overall membership ratio

The notion of overall membership is associated with the shape of a fuzzy number, regardless of where this shape is placed on the real line. To illustrate, consider two comparisons in **Figure 3**. In **Figure 3a**, while F_A and F_B have different representative x-values, their overall membership values should be the same due to the common shape. In contrast, F_C in **Figure 3b** should have higher overall membership than F_D as F_C's membership values are higher than or equal to those of F_D over the common support (note: the common support is not necessary; it just makes the comparison easier to observe).

To capture the above idea of the overall membership of a fuzzy number, we formulate the ratio using two areas: the shape's area and the full membership area over the same support. Also, we keep the concept of α-cut interval so that the decision maker can identify the minimum level of membership that is relevant for their ranking problem. **Figure 4** is used to illustrate the concept of both types of area. First, the shape's area is considered as the area under the fuzzy number and enclosed by the α-cut interval, as shaded by gray lines in **Figure 4**. Then, the full membership area is based on the rectangle with the width of the α-cut interval and the height of 1 (i.e., maximum membership). Accordingly, the shape's area (denoted as $area_{shape}$) and the full membership area (denoted as $area_{full}$) can be formulated as follows.

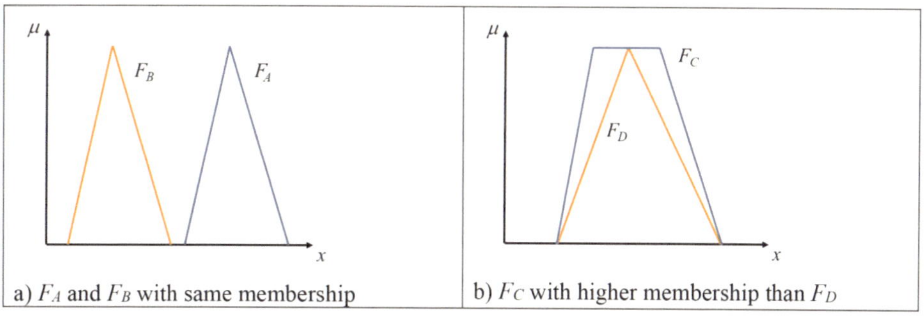

Figure 3.
Illustration of the concept for overall membership a) F_A *and* F_B *with same membership b)* F_C *with higher membership than* F_D.

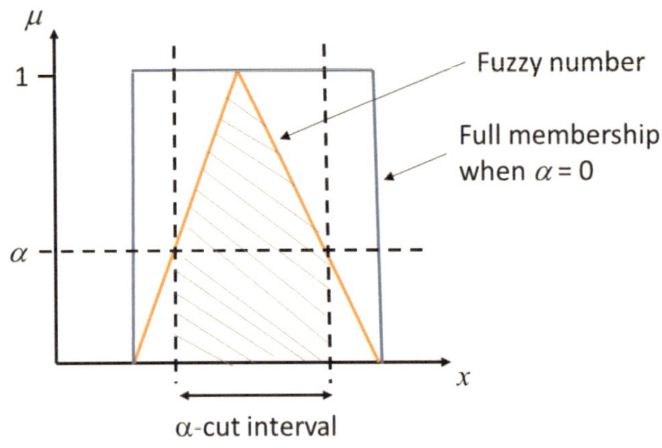

Figure 4.
Illustration of the shape's area and full membership area.

$$area_{shape}(F_A, \alpha) = \int_{l_{F_A}(\alpha)}^{r_{F_A}(\alpha)} \mu_{F_A}(x)dx \quad (9)$$

$$area_{full}(F_A, \alpha) = [r_{F_A}(\alpha) - l_{F_A}(\alpha)] \times 1 \quad (10)$$

The overall membership ratio of a fuzzy number (denoted as $mem(F_A, \alpha)$) can be expressed as follows.

$$mem(F_A, \alpha) = \frac{area_{shape}(F_A, \alpha)}{area_{full}(F_A, \alpha)} \quad (11)$$

Here, we suppose that higher overall membership ratio tends to yield a higher rank. We classify (overall) membership ratio as another secondary attribute because it may not be necessary for ranking problems with normal fuzzy numbers (e.g., if F_A is a normal triangular fuzzy number, $mem(F_A, 0)$ is always equal to 0.5). Yet, if this information is considered relevant, Section 5 will suggest one approach to use it as a modifying factor for ranking.

Notably, it is probably more common to apply two measures (instead of three) for FNR in literature (e.g., [value, ambiguity] and [average value, degree of deviation] as mentioned in Introduction). From there, they tend to integrate the information of range and membership ratio into one measure. We choose to handle such information in terms of two separate attributes for two reasons. First, the concepts of range and membership ratio are relatively direct for decision makers to visualize and interpret (thus supporting their intuition) in the comparison of fuzzy numbers. Second, range and membership ratio can indicate independent information. For example, consider two normal fuzzy numbers: one triangle and one trapezoid. While the trapezoid shape always yields a higher membership ratio, the ranges of both shapes can be changed arbitrarily, thus explaining the independence of range and membership ratio.

To demonstrate the evaluation of the three attributes, consider a fuzzy number: F_A = (1, 2, 3, 4; 1), which has a lower bound of 1 and an upper bound of 4. Its maximum membership value is 1, which covers the range between 2 and 3 (check **Figure 1** for an illustrative reference). Suppose that α = 0 (i.e., we consider the whole fuzzy number) and w = 1/12 (i.e., according to Abbasbandy and Hajjari [2]), we can evaluate the values of the three attributes according to the following:

- Representative x-value using Eq. (6): $rep(F_A, w)$ = (1/12) × 1 + (1/2–1/12) × 2 + (1/2–1/12) × 3 + (1/12) × 4 = 2.5
- X-value range using Eq. (8): $rng(F_A, \alpha) = r_{F_A}(\alpha) - l_{F_A}(\alpha)$ = 3
  - From Eq. (2): $l_{F_A}(\alpha)$ = 1 + (0/1) × (2–1) = 1
  - From Eq. (3): $r_{F_A}(\alpha)$ = 4 + (0/1) × (3–4) = 4
- Overall membership ratio using Eq. (11): $mem(F_A, \alpha) = \frac{area_{shape}(F_A, \alpha)}{area_{full}(F_A, \alpha)} = \frac{2}{3}$
  - From Eq. (9) = $area_{shape}(F_A, \alpha)$ = trapezoid's area = (1 + 3) × 1/2 = 2
  - From Eq. (10) = $area_{full}(F_A, \alpha) = [4–1] \times 1 = 3$

3.4 Pareto optimality

After defining three attributes, we can rank fuzzy numbers for some cases using the Pareto optimality principle [4]. In a less formal expression, we have $F_A \succcurlyeq F_B$ if $rep(F_A, w) \geq rep(F_B, w)$, $rng(F_A, \alpha) \leq rng(F_B, \alpha)$ and $mem(F_A, \alpha) \geq mem(F_B, \alpha)$. To examine how well these attributes can speak for the ranking intuition, numerical examples are used in the next sub-section to check the following situations.

- If two fuzzy numbers can be ranked based on Pareto optimality, this ranking order should be considered "obvious" to the ranking intuition with less room for arguments.
- If two fuzzy numbers cannot be ranked based on Pareto optimality, decision makers can effectively use the selected attribute to explain their arguments.

3.5 Numerical examples

The numerical cases from Bortolan and Degani [38] are employed for demonstration, and they can illustrate systematically how the selected attributes are changed with different fuzzy numbers. While we keep the case labels from Bortolan and Degani [38] for cross checking, we classify these cases into five groups for discussion. Also, we follow Abbasbandy and Hajjari [2] by setting w = 1/12 to evaluate $rep(F_A, w)$. Also we set α = 0 for $rng(F_A, \alpha)$ and $mem(F_A, \alpha)$ in this numerical demonstration.

Group 1: Non-overlapping, triangular fuzzy numbers

This group covers the cases of a, b, c, d and e from Bortolan and Degani [38], and the results are shown in **Table 1**. By examining the Pareto optimality with the three attributes, we can first pass the membership ratio because $mem(F_A, 0)$ is always equal to 0.5 if F_A is normal and triangular. The rankings of fuzzy numbers in cases a to d are obvious as the fuzzy numbers with higher representative x-values have the same (i.e., cases a, b, d) or smaller (i.e., case c) ranges. In case e, while the fuzzy number F_{E3} is ranked highest, we cannot immediately rank F_{E2} higher than F_{E1} based on Pareto optimality only since F_{E2} has a larger range. Through these five cases, we want to note

			Representative x-value (*rep*)	Range (*rng*)	Membership ratio (*mem*)
Case a	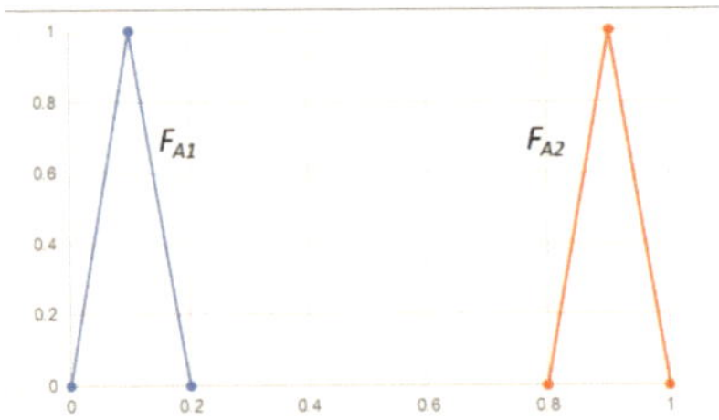	F_{A1}	0.1	0.2	0.5
		F_{A2}	0.9	0.2	0.5
Case b	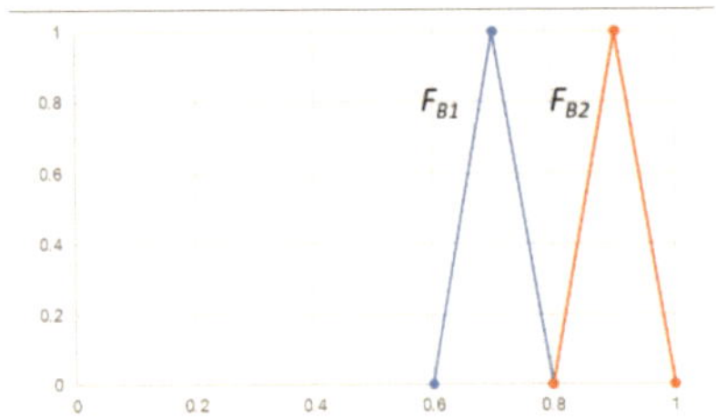	F_{B1}	0.7	0.2	0.5
		F_{B2}	0.9	0.2	0.5
Case c	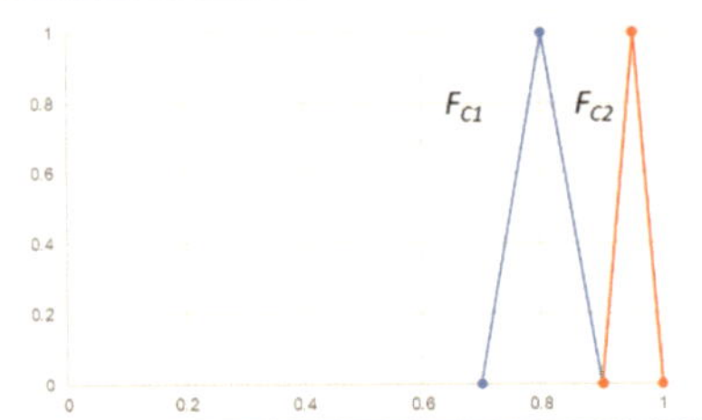	F_{C1}	0.8	0.2	0.5
		F_{C2}	0.95	0.1	0.5
Case d	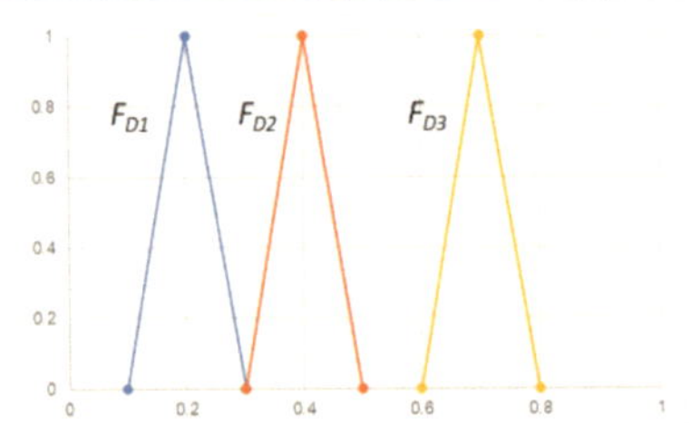	F_{D1}	0.2	0.2	0.5
		F_{D2}	0.4	0.2	0.5
		F_{D3}	0.7	0.2	0.5
Case e	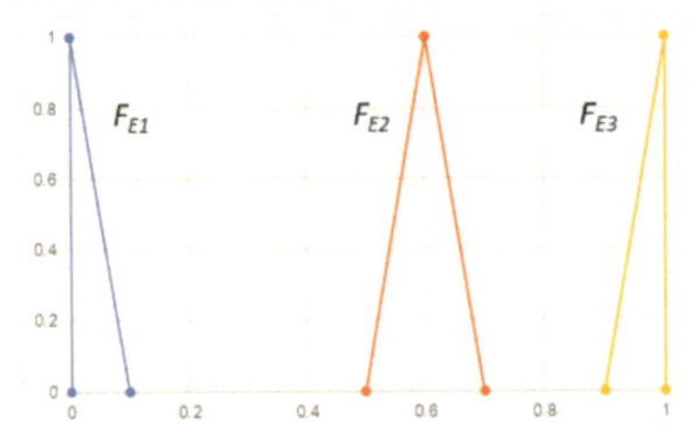	F_{E1}	0.0083	0.1	0.5
		F_{E2}	0.6	0.2	0.5
		F_{E3}	0.9917	0.1	0.5

Table 1.
Results of comparing non-overlapping, triangular fuzzy numbers.

that the proposed attributes vary according to our "intuition" to interpret and rank fuzzy numbers (e.g., check how representative x-values and ranges vary independently in these cases).

Group 2: Overlapping, triangular fuzzy numbers

This group covers the cases of f, i and l from Bortolan and Degani [38], and the results are shown in **Table 2**. In case f, while F_{F2} should be ranked higher than F_{F1} due to higher representative x-value shown in **Table 2**, we should note that this ranking is sensitive to the pre-set value of w. If $w < 1/6$ (i.e., more emphasis to the core values), we have $rep(F_{F2}) > rep(F_{F1})$. If $w \geq 1/6$ (i.e., more emphasis to the boundary values), we have $rep(F_{F1}) \geq rep(F_{F2})$.

In contrast, as the fuzzy numbers in case i share the same support, their ranking is not sensitive to the value of w. Finally, the ranking in case l depends on the information of range, and our intuition assumes that smaller range is better. Notably, our intuition here is not universal, and some decision maker can rank a fuzzy number of larger range higher for a positive likelihood of higher x-values. Here we are not arguing which "intuition" (or ranking rule) is right. Instead, we want to keep the intuition more transparent through explicit attributes so that researchers can argue their ranking intuitions on a common ground.

Group 3: Triangular and trapezoidal fuzzy numbers

This group covers the cases of g and h from [38], and the results are shown in **Table 3**. The trapezoid fuzzy numbers have a large shape, giving higher values of range and membership ratio. The triangular fuzzy numbers in both cases have higher

			Representative x-value (*rep*)	Range (*rng*)	Membership ratio (*mem*)
Case f	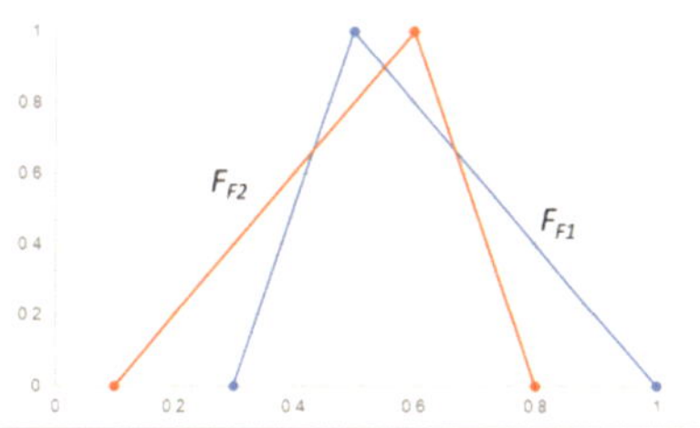	F_{F1}	0.525	0.7	0.5
		F_{F2}	0.575	0.7	0.5
Case i	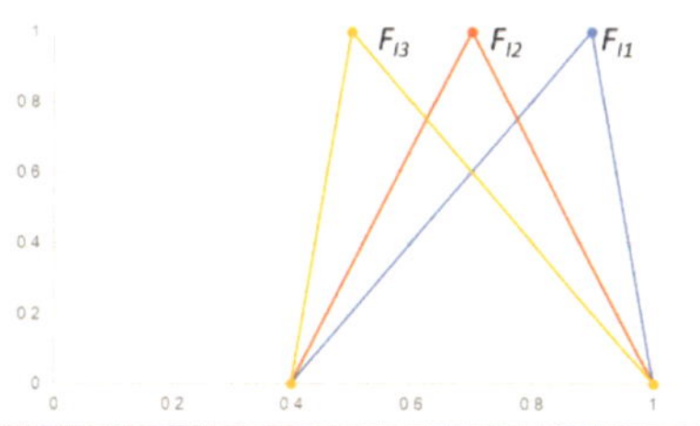	F_{I1}	0.8667	0.6	0.5
		F_{I2}	0.7	0.6	0.5
		F_{I3}	0.5333	0.6	0.5
Case l	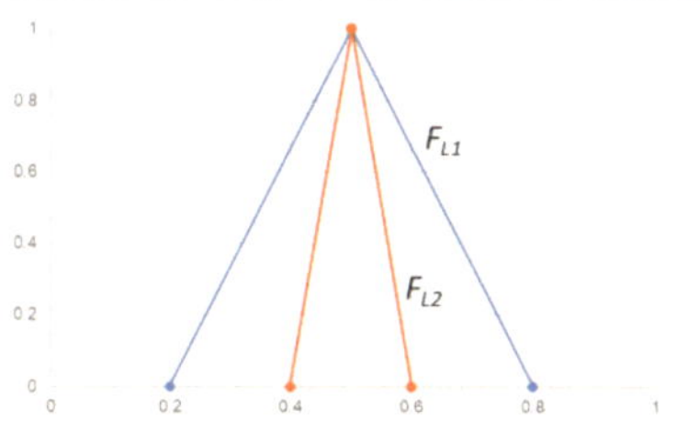	F_{L1}	0.5	0.6	0.5
		F_{L2}	0.5	0.2	0.5

Table 2.
Results of comparing overlapping, triangular fuzzy numbers.

			Representative x-value (*rep*)	**Range (*rng*)**	**Membership ratio (*mem*)**
Case g	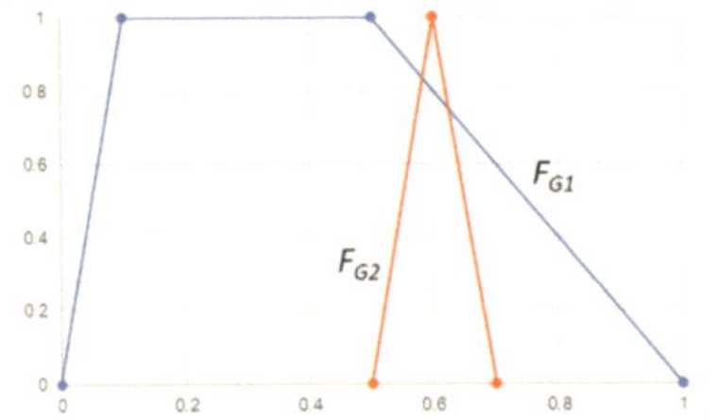	F_{G1}	0.3333	1	0.7
		F_{G2}	0.6	0.2	0.5
Case h	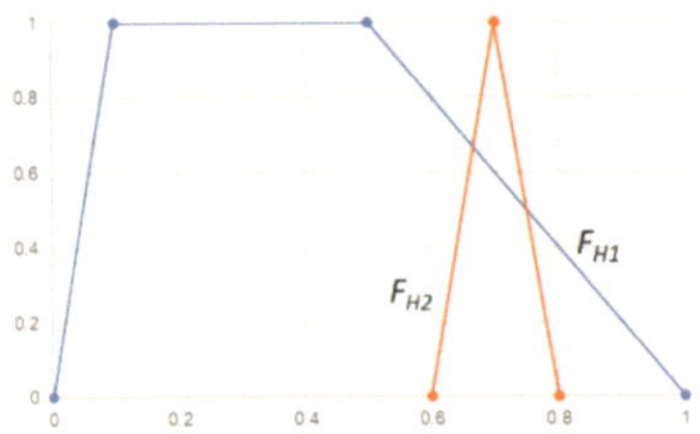	F_{H1}	0.3333	1	0.7
		F_{H2}	0.7	0.2	0.5

Table 3.
Results of comparing triangular and trapezoidal fuzzy numbers.

representative x-values. Their triangular shapes are the same, with a shift to the right side by 0.1 in case h. In view of Pareto optimality with three attributes, there is no dominant fuzzy number. Yet, we can note that if $F_{G2} \succcurlyeq F_{G1}$ in case g, we would have $F_{H2} \succcurlyeq F_{H1}$ in case h. It is because $F_{H2} \succcurlyeq F_{G2}$ due to Pareto optimality and $F_{G1} = F_{H1}$. This note should make sense when we observe the graphical shift of triangular fuzzy numbers from F_{G2} to F_{H2} in **Table 3**. This demonstrates how the three attributes can characterize some intuitive reasoning in FNR.

Group 4: Nested fuzzy numbers.

This group covers the cases of j and k from [38], and the results are shown in **Table 4**. In case j, F_{J2} is created by shifting the lower bound of F_{J1} to the left; F_{J2} and F_{J3} share the same support with a different shape. Fuzzy numbers in case k have a similar pattern in an opposite direction (see **Table 4**). By checking from the order $F_{J1} \rightarrow F_{J2} \rightarrow F_{J3}$ or $F_{K1} \rightarrow F_{K2} \rightarrow F_{K3}$, we argue that the three attributes can reasonably capture and quantify the characteristics of these fuzzy numbers.

			Representative x-value (*rep*)	**Range (*rng*)**	**Membership ratio (*mem*)**
Case j	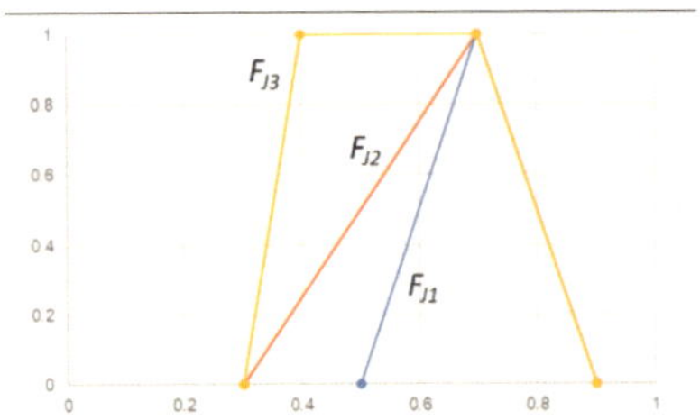	F_{J1}	0.7	0.4	0.5
		F_{J2}	0.6833	0.6	0.5
		F_{J3}	0.5583	0.6	0.75

			Representative x-value (*rep*)	Range (*rng*)	Membership ratio (*mem*)
Case k	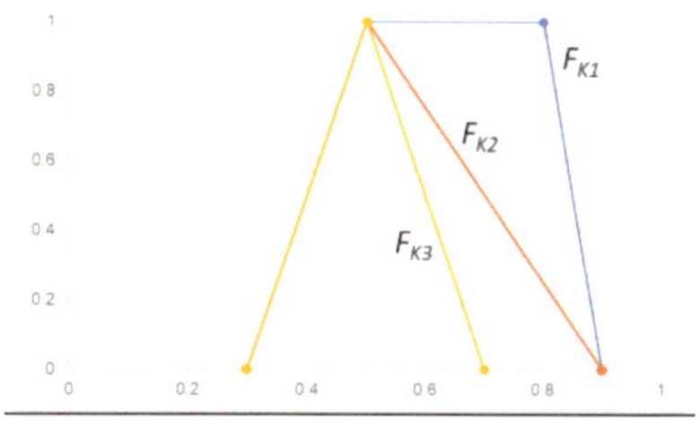	F_{K1}	0.6417	0.6	0.75
		F_{K2}	0.5167	0.6	0.5
		F_{K3}	0.5	0.4	0.5

Table 4.
Results of comparing nested fuzzy numbers.

Group 5: Non-normal fuzzy numbers.

Non-normal fuzzy numbers have their maximum membership less than 1 (i.e., $h_A < 1$). Notably, the literature of FNR often assumes normal fuzzy numbers (e.g., [28]). By inspecting the earlier cases, we should note that the variations of membership ratio of normal fuzzy numbers do not change much (from 0.5 for triangular to 0.7 or 0.75 for trapezoidal). Thus, it is not unreasonable if one chooses not to consider membership ratio for comparing normal fuzzy numbers. Yet, non-normal fuzzy numbers will open other possibilities, where the membership ratio can be an important consideration.

This group covers the cases of n, o, p, q and r from Bortolan and Degani [38], and the results are provided in **Table 5**. As shown in **Table 5**, the values of membership ratio vary more significantly as some fuzzy numbers have smaller maximum membership. Consequently, the trade-off consideration can be more challenging. For example, how should we compare F_{N1} and F_{N2} in case n with the trade-off of

			Representative x-value (*rep*)	Range (*rng*)	Membership ratio (*mem*)
Case n	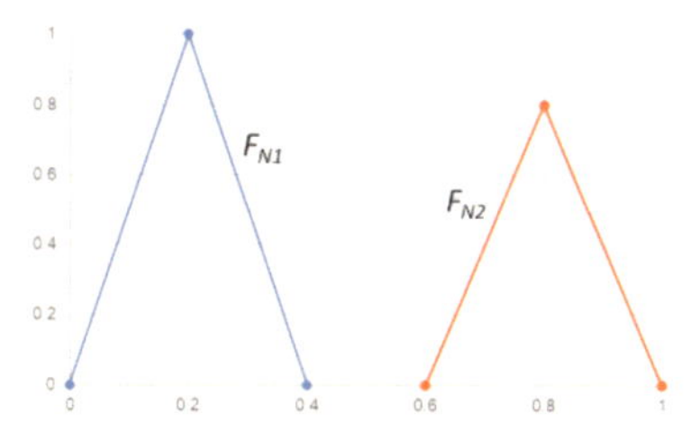	F_{N1}	0.2	0.4	0.5
		F_{N2}	0.8	0.4	0.4
Case o	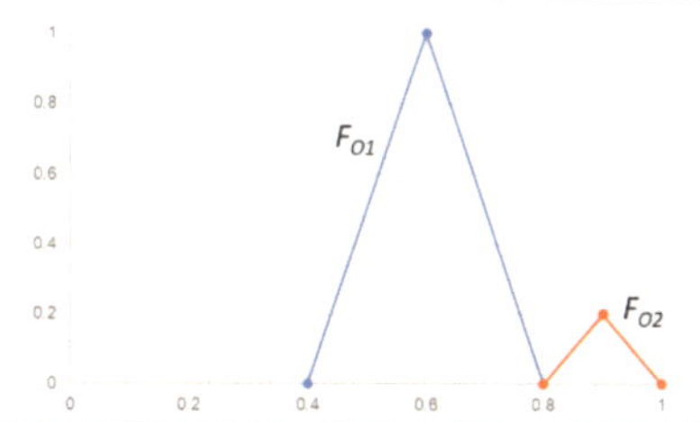	F_{O1}	0.6	0.4	0.5
		F_{O2}	0.9	0.2	0.1

			Representative x-value (*rep*)	**Range (*rng*)**	**Membership ratio (*mem*)**
Case p	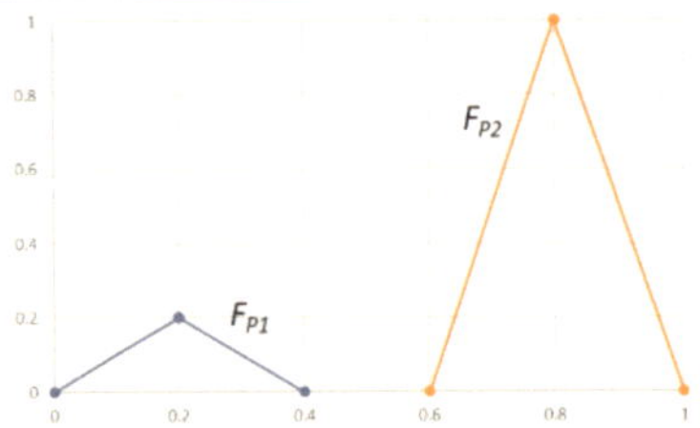	F_{P1}	0.2	0.4	0.1
		F_{P2}	0.8	0.4	0.5
Case q	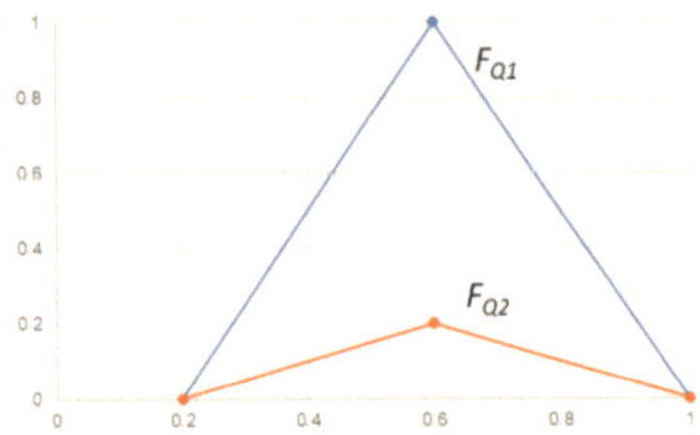	F_{Q1}	0.6	0.8	0.5
		F_{Q2}	0.6	0.8	0.1
Case r	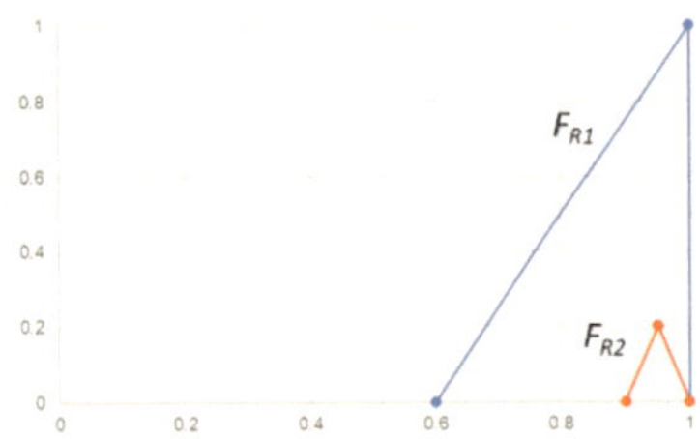	F_{R1}	0.9667	0.4	0.5
		F_{R2}	0.95	0.1	0.1

Table 5.
Results of comparing non-normal fuzzy numbers.

representative x-value and membership ratio (similarly for case o)? While we see $F_{P2} \succcurlyeq F_{P1}$ in case p and $F_{Q1} \succcurlyeq F_{Q2}$ in case q due to Pareto optimality, the trade-off consideration is present in case r with different values of range.

The main theme of this section is that we need some attributes to characterize our intuition for FNR. Otherwise, it is difficult to get a common ground for constructive arguments. In this section, we choose three attributes to make clear our "intuition" for FNR. Aligned with the note in Keeney and Raiffa [4], we do not claim the uniqueness of this selection of attributes for FNR. Other researchers can propose other sets of attributes to characterize their intuition.

4. Aggregation: proposal of a ranking index

If the Pareto optimality principle cannot rank two fuzzy numbers, trade-off consideration is required to finalize the ranking decision. That is, a fuzzy number of a higher rank must have some "weaker" aspect in terms of the three attributes but its "stronger" aspect is sufficient to bring it to a higher rank overall. This ranking process should involve an aggregation that combines all aspects into an overall evaluation and then determines the ranking result. This section will propose a ranking index for aggregation along with numerical examples.

4.1 Discount factors and ranking index

As discussed in Section 3, representative x-values are used as the primary attribute to rank fuzzy numbers. Then, we view the information of range and membership ratio as secondary attributes that will "discount" the representative x-values. To illustrate, consider a crisp number, 5, which has the representative x-value of 5, range of 0 and membership ratio of 1. If a fuzzy number with the representative x-value of 5 has a range larger than 0 and a membership ratio less than 1, this fuzzy number should be ranked lower than the crisp number 5. The discount factors are intended to capture this idea. Let $I_{rank}(F_A)$ be the index as the discounted representative x-value of F_A for ranking, and it can be formulated as follows.

$$I_{rank}(F_A) = d_{rng}(F_A) \cdot d_{mem}(F_A) \cdot rep(F_A, w) \tag{12}$$

where $d_{rng}(F_A)$ and $d_{mem}(F_A)$ are the discount factors associated with range and membership ratio, respectively. To quantify these discount factors, we consider the following conditions:

- $0 \leq d_{rng}(F_A) \leq 1$ and $0 \leq d_{mem}(F_A) \leq 1$
- If $rng(F_A, \alpha) \geq rng(F_B, \alpha)$, $d_{rng}(F_A) \leq d_{rng}(F_B)$.
- If $mem(F_A, \alpha) \geq mem(F_B, \alpha)$, $d_{mem}(F_A) \geq d_{mem}(F_B)$.

Apparently, many forms of formulations can be used for the discount factors and satisfy these conditions. In this chapter, we use a simple ratio with respect to some reference (or extreme) values. Let rng_{min} be the minimum reference for range, and mem_{max} be the maximum reference for membership ratio. We also set that $rng_{min} > 0$ and $0 < mem_{max} \leq 1$. Then, the discount factors for F_A can be formulated as follows.

$$d_{rng}(F_A) = \frac{rng_{min}}{rng(F_A, \alpha)} \tag{13}$$

$$d_{mem}(F_A) = \frac{mem(F_A, \alpha)}{mem_{max}} \tag{14}$$

With these discount factors, if F_A has a range equal to rng_{min}, its discount factor, $d_{rng}(F_A)$, is equal to 1 (i.e., no discount). A similar effect is also set for $d_{mem}(F_A)$. The selection of the values for rng_{min} and mem_{max} depends on how decision makers interpret the discount ratio for their ranking problems. One suggestion is to identify the minimum range and the maximum membership ratio from the set of fuzzy numbers to be ranked. That is, suppose that $\boldsymbol{FR} = \{F_A, F_B, F_C, \dots\}$ be the set of fuzzy numbers that need to be ranked in a problem. We can select rng_{min} and mem_{max} according to the following equations. Then, we can interpret the discount ratio with respect to the "best values" among the set of fuzzy numbers in the problem.

$$rng_{min} = min\,\{rng(F_A, \alpha), rng(F_B, \alpha), rng(F_C, \alpha) \dots\} \tag{15}$$

$$mem_{max} = max\,\{mem(F_A, \alpha), mem(F_B, \alpha), mem(F_C, \alpha) \dots\} \tag{16}$$

4.2 Overview of the ranking method

After defining the attributes in Section 3 and the ranking index in Section 4.1, this sub-section will overview our proposed approach to rank fuzzy numbers. The procedure to determine the ranking index is illustrated in **Figure 5**. Given a fuzzy number F_A, we first determine the values of three attributes: representative x-value, x-value range and overall membership ratio. Then, we can evaluate the discount factors for x-value range and overall membership ratio. In the end, we can determine the ranking index for the given fuzzy number.

Suppose that we are tasked to rank a set of fuzzy numbers. We first determine the ranking index for each fuzzy number. Then, we can use the index, I_{rank}, for this ranking task. That is, if $I_{rank}(F_A) \geq I_{rank}(F_B)$, we rank F_A higher than F_B, symbolically, $F_A \succsim F_B$.

4.3 Numerical examples

As a recall from Section 3, we set w = 1/12 and α = 0 to evaluate representative x-value, range and membership ratio. We use Eqs. (15) and (16) to obtain rng_{min} and mem_{max} and then calculate the values of the discounts and the ranking index. We reuse the numerical examples from Section 3.5 with the cases where Pareto optimality cannot finalize the ranking. The results are presented in **Table 6**.

Case e comes from Group 1 (see **Table 1**), where F_{E3} is ranked on the top per Pareto optimality (same result from the ranking index). Between F_{E1} and F_{E2}, though F_{E2} should be ranked higher intuitively, trade-off is involved logically because F_{E2} has a large range (reflected in its range discount of 0.5 as well). Per the ranking index, we still have $F_{E2} \succsim F_{E1}$, which matches the general intuition.

Cases g and h come from Group 3 (see **Table 3**), where wide trapezoidal fuzzy numbers are compared with narrow triangular fuzzy numbers. Per the ranking index, the triangular fuzzy numbers are ranked higher mainly because of the large difference of the range discount (1 vs. 0.2). In contrast, the difference of the membership ratio discount is less substantial.

Cases j and k come from Group 4 (see **Table 4**), where two triangular fuzzy numbers are nested in a trapezoidal fuzzy number. In both cases, the wider triangular fuzzy numbers (i.e., F_{J2} and F_{K2}) are ranked lowest as they receive both discounts

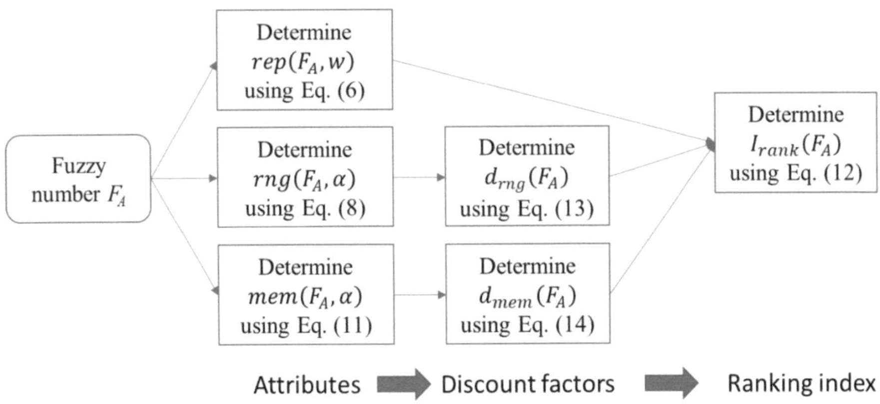

Figure 5.
Procedure to determine the ranking index.

	Fuzzy number	Range discount (d_{rng})	Mem. discount (d_{mem})	Ranking index (I_{rank})
Case e	F_{E1} = (0, 0, 0, 0.1; 1)	1	1	0.0083
	F_{E2} = (0.5, 0.6, 0.6, 0.7; 1)	0.5	1	0.3
	F_{E3} = (0.9, 1, 1, 1; 1)	1	1	0.9917
Case g	F_{G1} = (0, 0.1, 0.5, 1; 1)	0.2	1	0.0667
	F_{G2} = (0.5, 0.6, 0.6, 0.7; 1)	1	0.7143	0.4286
Case h	F_{H1} = (0, 0.1, 0.5, 1; 1)	0.2	1	0.0667
	F_{H2} = (0.6, 0.7, 0.7, 0.8; 1)	1	0.7143	0.5
Case j	F_{J1} = (0.5, 0.7, 0.7, 0.9; 1)	1	0.6667	0.4667
	F_{J2} = (0.3; 0.7, 0.7, 0.9; 1)	0.6667	0.6667	0.3037
	F_{J3} = (0.3, 0.4, 0.7, 0.9; 1)	0.6667	1	0.3722
Case k	F_{K1} = (0.3, 0.5, 0.8, 0.9; 1)	0.6667	1	0.4278
	F_{K2} = (0.3, 0.5, 0.5, 0.9; 1)	0.6667	0.6667	0.2296
	F_{K3} = (0.3, 0.5, 0.5, 0.7; 1)	1	0.6667	0.3333
Case n	F_{N1} = (0, 0.2, 0.2, 0.4; 1)	1	1	0.2
	F_{N2} = (0.6, 0.8, 0.8, 1; 0.8)	1	0.8	0.64
Case o	F_{O1} = (0.4, 0.6, 0.6, 0.8; 1)	0.5	1	0.3
	F_{O2} = (0.8, 0.9, 0.9, 1; 0.2)	1	0.2	0.18
Case r	F_{R1} = (0.6, 1, 1, 1; 1)	0.25	1	0.2417
	F_{R2} = (0.9, 0.95, 0.95, 1; 0.2)	1	0.2	0.19

Table 6.
Ranking index results for cases with trade-off consideration.

(i.e., $d_{rng} = d_{mem} = 0.6667$). In case j, the narrower triangular fuzzy number (F_{J1}) is ranked first because its representative x-value and range can "win" over its weaker membership ratio as compared to the trapezoidal fuzzy number (F_{J3}). In contrast, in case k, the trapezoidal fuzzy number (F_{K1}) "wins" because it has better representative x-value and membership ratio as compared to F_{K3}.

Cases n, o and r come from Group 5 (see **Table 5**). In case n, we have $F_{N2} \succsim F_{N1}$, as F_{N2} has higher representative x-value despite lower membership ratio (associated with the discount $d_{mem}(F_{N2}) = 0.8$). In cases o, we have $F_{O1} \succsim F_{O2}$ because the membership ratio of F_{O2} is substantially lower despite its higher representative x-value. In case r, the trade-off between range and membership ratio is relatively close. In the end, we have $F_{R1} \succsim F_{R2}$, as F_{R2} has a lower value of the discount from membership ratio (i.e., $d_{mem}(F_{R2}) = 0.2$ vs. $d_{rng}(F_{R1}) = 0.25$).

Notably, the judgment for ranking with trade-off can become difficult when the trade-off among the three attributes is getting close. We argue that such difficulty is fundamentally embedded into the problem structure of FNR, which involves multiple dimensions of considerations. Thus, our solution strategy is not about providing the best ranking procedure. Instead, we emphasize the importance of defining attributes to quantify the "intuition". Then, decision makers can explicitly explain their trade-off considerations in the ranking process.

5. Multi-attribute ranking method in practice

5.1 General suggestions for application

As we consider that ranking methods should be dependent on a given set of fuzzy numbers to be ranked (i.e., context-dependent), we want to discuss two types of adjustable elements of our proposed method in practice. The first type is the selection of attributes. Among three attributes: representative x-value, range and membership ratio, representative x-value should be a default choice as it intuitively corresponds to the ranking of real numbers (e.g., compare representative x-values of different fuzzy numbers). We suggest the class of ranking indices by Ban and Coroianu [28] (i.e., in Eq. (6)) as it satisfies the six axioms. To determine the weight, w, of this attribute, decision makers may consider sensitivity analysis for their given set of fuzzy numbers (i.e., how sensitive of the value of w can alter the ranking of two fuzzy numbers).

In contrast to representative x-value, the choice of range and membership ratio is optional. The attribute of x-value range is common in literature, and other formulations of this attribute (e.g., ambiguity and deviation degree as mentioned in Section 3.2) can be considered as a choice by decision makers for this attribute. If the ranking problem has fuzzy numbers of similar ranges (e.g., triangular fuzzy numbers with similar supports), we think it is legitimate not to consider range in FNR (in order to preserve some axiomatic properties, to be discussed in Section 6). The attribute of membership ratio is less common, and it should be more relevant for non-normal fuzzy numbers (e.g., normal triangular fuzzy numbers always have the same membership ratio equal to 0.5).

The second type of adjustable elements of our proposed method is the specification of the reference values (i.e., rng_{min} and mem_{max}) and the formulations of the discount factors. Notably, our formulations of discount factors (i.e., Eqs. (13) and (14)) are only one simple suggestion. One possible disadvantage of our discount factors is that they can be too sensitive to the reference values. For example, if two fuzzy numbers with the range values of 0.5 and 1 are compared, the range discount (d_{rng}) can be equal to 0.5 for one fuzzy number, cutting half of its representative x-value. Decision makers can consider adjusting the effects of discount factors through other formulations for their problems (e.g., additional scaling component).

5.2 Applicability to specific forms

The origin of fuzzy numbers can be viewed as a generalization of crisp numbers to describe approximate information. Consider the 5-tuple definition of a fuzzy number, $F_A = (a_1, a_2, a_3, a_4; h_A)$ as the generalized form of fuzzy numbers in this work. Accordingly, three specific forms can be considered as follows.

- Interval: if $a_1 = a_2, a_3 = a_4$ and $h_A = 1$;
- Ordered pair (an element of a fuzzy set): if $a_1 = a_2 = a_3 = a_4$ and $h_A < 1$;
- Crisp number: if $a_1 = a_2 = a_3 = a_4$ and $h_A = 1$.

Then, we want to investigate the reducibility property that whether a ranking method can still be applicable if the above specific forms are considered. **Table 7** shows that our proposed ranking method can be still used for these specific forms. First, a general fuzzy number can be evaluated using the equations as listed in the first

	Representative x-value (*rep*)	Range (*rng*)	Membership ratio (*mem*)	Range discount (d_{rng})	Membership discount (d_{mem})
Fuzzy number	Eq. (6)	Eq. (8)	Eq. (11)	Eq. (13)	Eq. (14)
Interval	$(a_2 + a_3)/2$	$(a_4 – a_1)$	1	Eq. (13)	1
Ordered pair	a_1 $(= a_2 = a_3 = a_4)$	0	h_A	1	Eq. (14)
Crisp number	a_1 $(= a_2 = a_3 = a_4)$	0	1	1	1

Table 7.
Overview of the reducibility property.

row of **Table 7**. When these equations are applied to interval, ordered pair and crisp number, we can obtain the results that match our expectations. For example, the representative x-value of an interval will be the midpoint of a_2 and a_3, and a crisp number has no discount effect (i.e., $d_{rng} = 1$ and $d_{mem} = 1$). In this way, our proposed method can be used to compare fuzzy numbers with intervals or crisp numbers in the same methodical framework.

6. Information content and axiomatic properties

Since the pioneer work by Wang and Kerre [1], researchers have examined the axiomatic properties (i.e., the six axioms listed in Section 2) of fuzzy number ranking methods. The intent of this section is to discuss how the information content for ranking can influence the axiomatic properties in the context of our ranking approach. One key message is that the satisfaction of axioms depends on the selection of information that is deemed relevant to FNR. If more information is selected and considered for ranking, the ranking method is more likely to violate the axioms. This message is aligned with the topic of information basis in the analysis of the Arrow's Impossibility Theorem [39, 40].

6.1 Analysis of Axiom 4

Axiom 4 somewhat dictates the ranking of non-overlapping fuzzy numbers. If we only consider representative x-values for ranking (i.e., no range, membership ratio and discount factors), our ranking procedure will directly follow the results from Ban and Coroianu [28], and it will thus satisfy Axiom 4. However, if range and membership ratio are considered as relevant information for ranking, Axiom 4 can be violated, and the reason is given below.

Axiom 4 only focuses on the boundary values without considering any distributional information (e.g., range and membership). When multiple attributes are considered for ranking, Axiom 4 can be violated by strengthening the distributional aspect of the inferior fuzzy number (in view of Axiom 4). For example, we have $F_{O2} \succsim F_{O1}$ in case o (see **Table 5**) according to Axiom 4, no matter how small of membership ratio of F_{O2}. However, when we consider range and membership ratio, we obtain $F_{O1} \succsim F_{O2}$ (see **Table 6**), which violates Axiom 4. Notably, the logic of such violation can be held whenever we deem membership ratio as relevant information for FNR, regardless of the details of the ranking procedures.

Notably, this discussion is not about rejecting Axiom 4. Instead, we want to explain one logical tension with Axiom 4. That is, Axiom 4 dictates some ranking of fuzzy numbers based on their boundary values only, and this opens a chance for the information of range and membership ratio to violate Axiom 4. Alternately, if we choose the index class by [28], representative x-values will be the only information considered for ranking, and the information of range (for example) will become irrelevant for FNR. In other words, if we consider that FNR should involve trade-off with multiple attributes in addition to representative x-value, Axiom 4 could be violated in some situations.

6.2 Dependence of rng_{min} and mem_{max}

As we use reference values (i.e., rng_{min} and mem_{max}) to evaluate the discount factors (i.e., d_{rng} and d_{mem}), the ranking index, I_{rank}, belongs to the second class of ranking indices according to the classification by [1]. In their axiomatic analysis, they have identified five indices of the second class, i.e., $J^K()$, $K()$, $CH^K()$, $W()$ and $KP^K()$, which all do not satisfy Axioms 5 and 6. Without listing counter-examples, I_{rank} of the same class follows the same conclusion because they share a common feature of these ranking indices, i.e., use of reference values.

Why using reference values could violate Axioms 5 and 6? It is because the index values would depend on the information that is external to the fuzzy numbers themselves. For example, if a fuzzy number with a very small range is added to a set of fuzzy numbers for ranking (i.e., $\boldsymbol{FR}$), this newly added fuzzy number will decrease the reference value, rng_{min}, and thus generally decrease the range discount values (d_{rng}) for the original set of fuzzy numbers. Then, all values of I_{rank} would change because of adding a new fuzzy number to the set (i.e., $\boldsymbol{FR}$).

While it seems undesirable by setting rng_{min} and mem_{max} per individual sets of fuzzy numbers, can we simply set these two values as universal numbers that are applicable to all ranking problems (e.g., simply set $mem_{max} = 1$)? Theoretically, it is a viable option. However, by doing so, we somehow lose our interpretation of "discount" factors that are relevant to a given set of fuzzy numbers that we want to rank in the problem. For example, it is not easy to interpret if the range of a fuzzy number, say 5, is large or small until we know a reference for comparison (e.g., if $rng_{min} = 1$, the range of 5 will quite large). In other words, the reference values, rng_{min} and mem_{max}, provide a numerical context as relevant information for comparison.

To close this section, we want to make a note about the historical development of the Arrow's Impossibility Theorem, which proves that no voting method (or social welfare function) can satisfy a set of "reasonable" properties (or axioms) [41]. One famous "escaping route" is the information basis approach, which classifies the information content (or availability) for interpersonal comparisons with different axiomatic results [39, 40]. Back to our context, if representative x-value is taken as the only relevant information for FNR, the results by [28] are sufficient to design a ranking index that satisfies the six axioms in Section 2. However, if additional information is considered for FNR, the axiomatic properties cannot be guaranteed. To us, this tension seems fundamental.

7. Conclusion

The main theme of this chapter is to use the multi-attribute approach to analyze and address the problems of fuzzy number ranking (FNR) since numerous ranking methods in literature have implicated multiple attributes in their ranking intuition. The multi-attribute approach has two phases: the selection of attributes and the formulation of the aggregation function. The selection of attributes determines what information is deemed relevant for FNR, and the aggregation function controls the trade-off of the attribute values of fuzzy numbers. In this work, we propose three attributes (i.e., representative x-value, range and membership ratio) as three possible dimensions to evaluate a fuzzy number. In aggregation, we formulate the discount factors for range and membership ratio to modify the representative x-value of a fuzzy number for FNR. The proposed method has been illustrated via numerical examples to reveal the rationale of using multiple attributes to articulate the intuition behind FNR.

In future work, there can be two directions to consider: practice and theory. In the practice direction, we can develop more methodical guidance toward the selection and formulations of attributes and the aggregation procedures. In particular, we can formalize the ranking intuition in terms of the selected attributes and the trade-off rationale through the aggregation approach for different FNR problems. Along this effort, we can also compare the ranking results from this multi-attribute approach with other FNR approaches. In the theory direction, while this chapter has initially explored the tension between information content and the satisfaction of the FNR axioms. This tension should call for more mathematical analyses such as classification of information content for FNR and relaxation of axioms for expanded information basis.

Author details

Simon Li
Department of Mechanical and Manufacturing Engineering, Schulich School of Engineering, University of Calgary, Alberta, Canada

*Address all correspondence to: simoli@ucalgary.ca

References

[1] Wang X, Kerre EE. Reasonable properties for the ordering of fuzzy quantities (I). Fuzzy Sets and Systems. 2001;**118**:375-385

[2] Abbasbandy S, Hajjari T. A new approach for ranking of trapezoidal fuzzy numbers. Computers & Mathematics with Applications. 2009;**57**: 413-419. DOI: 10.1016/j.camwa.2008.10.090

[3] Chou S-Y, Dat LQ, Yu VF. A revised method for ranking fuzzy numbers using maximizing set and minimizing set. Computers & Industrial Engineering. 2011;**61**:1342-1348. DOI: 10.1016/j.cie.2011.08.009

[4] Keeney RL, Raiffa H. Decisions with Multiple Objectives: Preferences and Value Tradeoffs. New York: John Wiley & Sons; 1976

[5] Chen S-H. Ranking fuzzy numbers with maximizing set and minimizing set. Fuzzy Sets and Systems. 1985;**17**:113-129

[6] Wang Y-M, Luo Y. Area ranking of fuzzy numbers based on positive and negative ideal points. Computers & Mathematics with Applications. 2009;**58**: 1769-1779. DOI: 10.1016/j.camwa.2009.07.064

[7] Murakami S, Maeda H, Imamura S. Fuzzy decision analysis on the development of centralized regional energy control system. In: IFAC Proceedings Volumes, IFAC Symposium on Fuzzy Information, Knowledge Representation and Decision Analysis, Marseille, France, 19–21 July 1983. Vol. 16. Elsevier; 1983. pp. 363-368. DOI: 10.1016/S1474-6670(17)62060-3. Available from: https://www.sciencedirect.com/journal/ifac-proceedings-volumes/vol/16/issue/13

[8] Wang Y-J, Lee H-S. The revised method of ranking fuzzy numbers with an area between the centroid and original points. Computers & Mathematics with Applications. 2008;**55**:2033-2042. DOI: 10.1016/j.camwa.2007.07.015

[9] Delgado M, Vila MA, Voxman W. On a canonical representation of fuzzy numbers. Fuzzy Sets and Systems. 1998; **93**:125-135. DOI: 10.1016/S0165-0114(96)00144-3

[10] Bodjanova S. Median value and median interval of a fuzzy number. Information Sciences. 2005;**172**:73-89. DOI: 10.1016/j.ins.2004.07.018

[11] Asady B. Revision of distance minimization method for ranking of fuzzy numbers. Applied Mathematical Modelling. 2011;**35**:1306-1313. DOI: 10.1016/j.apm.2010.09.007

[12] Ezzati R, Allahviranloo T, Khezerloo S, Khezerloo M. An approach for ranking of fuzzy numbers. Expert Systems with Applications. 2012;**39**: 690-695. DOI: 10.1016/j.eswa.2011.07.060

[13] Ban A, Brândaş A, Coroianu L, Negruţiu C, Nica O. Approximations of fuzzy numbers by trapezoidal fuzzy numbers preserving the ambiguity and value. Computers & Mathematics with Applications. 2011;**61**:1379-1401. DOI: 10.1016/j.camwa.2011.01.005

[14] Chutia R, Chutia B. A new method of ranking parametric form of fuzzy numbers using value and ambiguity. Applied Soft Computing. 2017;**52**: 1154-1168. DOI: 10.1016/j.asoc.2016.09.013

[15] Zhu L, Xu R. Ranking fuzzy numbers based on fuzzy mean and standard

deviation. In: 2011 Eighth International Conference on Fuzzy Systems and Knowledge Discovery (FSKD). Presented at the 2011 Eighth International Conference on Fuzzy Systems and Knowledge Discovery (FSKD 2011). Shanghai: IEEE; 2011. pp. 854-857. DOI: 10.1109/FSKD.2011.6019703

[16] Gu Q, Xuan Z. A new approach for ranking fuzzy numbers based on possibility theory. Journal of Computational and Applied Mathematics. 2017;**309**:674-682. DOI: 10.1016/j.cam.2016.05.017

[17] Chutia R. Ranking of fuzzy numbers by using value and angle in the epsilon-deviation degree method. Applied Soft Computing. 2017;**60**:706-721. DOI: 10.1016/j.asoc.2017.07.025

[18] Yu VF, Chi HTX, Dat LQ, Phuc PNK, Shen C. Ranking generalized fuzzy numbers in fuzzy decision making based on the left and right transfer coefficients and areas. Applied Mathematical Modelling. 2013;**37**:8106-8117. DOI: 10.1016/j.apm.2013.03.022

[19] Dinagar DS, Kamalanathan S. A method for ranking of fuzzy numbers using new area method. International Journal of Fuzzy Mathematical Archive. 2015;**9**:61-71

[20] Jiang W, Luo Y, Qin X-Y, Zhan J. An improved method to rank generalized fuzzy numbers with different left heights and right heights. Journal of Intelligent & Fuzzy Systems. 2015;**28**:2343-2355. DOI: 10.3233/IFS-151639

[21] Thorani YLP, Rao PPB, Shankar NR. Ordering generalized trapezoidal fuzzy numbers using orthocentre of centroids. International Journal of Algebra. 2012;**6**:1069-1085

[22] Allahviranloo T, Jahantigh MA, Hajighasemi S. A new distance measure and ranking method for generalized trapezoidal fuzzy numbers. Mathematical Problems in Engineering. 2013;**2013**:1-6. DOI: 10.1155/2013/623757

[23] Chi HTX, Yu VF. Ranking generalized fuzzy numbers based on centroid and rank index. Applied Soft Computing. 2018;**68**:283-292. DOI: 10.1016/j.asoc.2018.03.050

[24] Rao PPB, Shankar NR. Ranking generalized fuzzy numbers using area, mode, spreads and weights. International Journal of Applied Science and Engineering. 2012;**10**:41-57

[25] Rao PPB, Shankar NR. Ranking fuzzy numbers with a distance method using circumcenter of centroids and an index of modality. Advances in Fuzzy Systems. 2011;**2011**:1-7. DOI: 10.1155/2011/178308

[26] Cheng C-H. A new approach for ranking fuzzy numbers by distance method. Fuzzy Sets and Systems. 1998;**95**:307-317. DOI: 10.1016/S0165-0114(96)00272-2

[27] Chu T-C, Tsao C-T. Ranking fuzzy numbers with an area between the centroid point and original point. Computers & Mathematics with Applications. 2002;**43**:111-117. DOI: 10.1016/S0898-1221(01)00277-2

[28] Ban A, Coroianu L. Simplifying the search for effective ranking of fuzzy numbers. IEEE Transactions on Fuzzy Systems. 2015;**23**:327-339. DOI: 10.1109/TFUZZ.2014.2312204

[29] Abbasi Shureshjani R, Darehmiraki M. A new parametric method for ranking fuzzy numbers. Indagationes Mathematicae. 2013;**24**:

518-529. DOI: 10.1016/j.indag.2013.02.002

[30] Saeidifar A. Application of weighting functions to the ranking of fuzzy numbers. Computers & Mathematics with Applications. 2011;**62**: 2246-2258. DOI: 10.1016/j.camwa.2011.07.012

[31] Kumar A, Singh P, Kaur P, Kaur A. RM approach for ranking of L–R type generalized fuzzy numbers. Soft Computing. 2011;**15**:1373-1381. DOI: 10.1007/s00500-010-0676-x

[32] Dombi J, Jónás T. Ranking trapezoidal fuzzy numbers using a parametric relation pair. Fuzzy Sets and Systems. 2020;**399**:20-43. DOI: 10.1016/j.fss.2020.04.014

[33] Van Hop N. Ranking fuzzy numbers based on relative positions and shape characteristics. Expert Systems with Applications. 2022;**191**:116312. DOI: 10.1016/j.eswa.2021.116312

[34] Dubois D, Prade H. Operations on fuzzy numbers. International Journal of Systems Science. 1978;**9**:613-626. DOI: 10.1080/00207727808941724

[35] Heilpern S. The expected value of a fuzzy number. Fuzzy Sets and Systems. 1992;**47**:81-86

[36] Wang Y-M, Yang J-B, Xu D-L, Chin K-S. On the centroids of fuzzy numbers. Fuzzy Sets and Systems. 2006; **157**:919-926. DOI: 10.1016/j.fss.2005.11.006

[37] Nasseri SH, Zadeh MM, Kardoost M, Behmanesh E. Ranking fuzzy quantities based on the angle of the reference functions. Applied Mathematical Modelling. 2013;**37**: 9230-9241. DOI: 10.1016/j.apm.2013.04.002

[38] Bortolan G, Degani R. A review of some methods for ranking fuzzy subsets. Fuzzy Sets and Systems. 1985;**15**:1-19

[39] Roberts KWS. Interpersonal comparability and social choice theory. The Review of Economic Studies. 1980; **47**:421. DOI: 10.2307/2297002

[40] Sen A. The possibility of social choice. The American Economic Review. 1999;**89**:349-378. DOI: 10.1257/aer.89.3.349

[41] Arrow KJ. Social Choice and Individual Values. 2nd ed. New Haven: Yale University Press; 1963

Chapter 3

Computing the Performance Parameters of the Markovian Queueing System FM/FM/1 In Transient State

Rostin Mabela Makengo Matendo, Jean Alonge W'Omatete, Herman Matondo Mananga, Jean Pierre Mukeba Kanyinda and Baudouin Adia Leti Mawa

Abstract

In this chapter, we utilize L–R method to calculate the parameters of performance of fuzzy Markovian queueing system FM/FM/1 in transient regime. The technique of calculating used is the arithmetic of L–R fuzzy numbers restricted to secant approximations. The membership function has helped us to represent graphically the curves of fuzzy parameters of performance into the space in three dimensions in transient regime in fuzzy environment. An illustrative example is given in the medical field to show the relevance of this study in operational research and in particular in queueing systems.

Keywords: performance parameters, fuzzy Markovian queueing systems, transient state, L–R method, computing

1. Introduction

Nowadays, in many areas of life, whether it is computer systems, communication systems, production systems, medical or health systems or any other system of daily life, the world is looking for the best quality of service and performance of systems.

Thus, Baynat [1] points out that "it is becoming inconceivable to build any system without first doing a performance analysis. This analysis is related to the prior knowledge of the performance parameters of the queueiting systems such as the average number of customers in the queue and in the system; as well as the average waiting time of customers in the queue and in the system."

The fundamental question posed in this chapter is: "Would the L–R method be able to compute the performance parameters of the fuzzy Markovian queueing system FM/FM/1 in transient regime?"

In the literature browsed in fuzzy mathematics, it is well shown that fuzzy queues are widely studied in steady state by Ning and Zhao [2], Ritha and Robert [3], Ritha and Menon [4].

Many researches on Markovian fuzzy queueing systems and scientific papers have been based on computing the performance parameters of Markovian fuzzy queueing systems in steady state by the method of sluggish alpha-cuts and the L–R method, see for example Li and Lee [5, 6]; Kao et al. [7]; Palpandi and Geetharamani [8]; Wang et al. [9]. But, the calculation of these parameters of the system under study in transient regime is a major preoccupation of operational researchers nowadays.

In this chapter, the novelty of our study is due to the fact that we have computed the performance parameters of the queueing system, in the transient regime and in a fuzzy environment where these parameters are time-dependent fuzzy functions, whereas for all the authors presented above, the performance parameters have been analyzed in steady state where the results obtained are real numbers.

L–R Fuzzy Mathematics plays an important role if it is widened to secant approximation. This arithmetic conducts to the same results as those obtained by the most used and well-known alpha-cut and interval arithmetic (see Mukeba; Dubois D. and Prade [10–12]).

To achieve this, our approach is broken down as follows: The second section will present the classical M/M/1 queueing model and give the performance parameters of the model in a transient state The third section will recall the notions of fuzzy set, fuzzy numbers, fuzzy number of L–R type, arithmetic of fuzzy numbers of L–R type and triangular fuzzy number. The fourth section will be devoted to fuzzy functions. The fifth section will give a description of the L–R method and the calculation procedure. The sixth section will deal with a numerical example that uses the L–R method in transient state. The seventh section will give the conclusion of this chapter.

2. Presentation of the queueing model M/M/1 et some performance parameters in transient state

2.1 Classical M/M/1 queueing system

Definition 1: A queueing system or M/M/1 queue is a Markovian process unfolding in L.D.P.(Life and Death Process) with birth and death rates defined respectively by:

$$\begin{cases} \lambda_n = \lambda, \lambda > 0 \\ \mu_n = \begin{cases} \mu & \text{if } n > 0 \\ 0 & \text{if } n = 0 \end{cases} \end{cases} \tag{1}$$

2.1.1 Assumptions (or characteristics) of the model

- Customer arrivals follow a Poisson distribution of parameter λ;
- Service times follow an expo-negative distribution of parameter μ;

- The system has a single server;
- Waiting room capacity is infinite;
- The system capacity is infinite;
- The service discipline is FIFO (first in first out) or PAPS (first come, first served).

2.2 Performance parameters of the M/M/1 queue in transient state

In the literature, it is well-known that a queue is stable if and only if (cfr. [13]): $\lambda < \mu$

- This condition makes it possible to determine the following performance parameters in transient state:

$$\tilde{N}_S(t) = \frac{\rho}{1-\rho}\left(1 - e^{-(\mu-\lambda)t}\right) \tag{2}$$

$$\tilde{T}_S(t) = \frac{\rho\left(1 - e^{-(\mu-\lambda)t}\right)}{\mu(1-\rho)\left[\rho + (1-\rho)e^{-(\mu-\lambda)t}\right]} \tag{3}$$

where $\tilde{N}_S(t)$ and $\tilde{T}_S(t)$ are, respectively, the average number of customers in the system and the average waiting time in the system at time $t(t \geq 0)$..

3. Fuzzy sets, fuzzy numbers, alpha-cuts and interval arithmetic, L–R fuzzy numbers, arithmetic of L–R fuzzy numbers, triangular fuzzy numbers

3.1 Fuzzy sets

Definition 2: Let E be a classical set or a universe. A fuzzy subset $\tilde{A}$ (or a fuzzy set $\tilde{A}$) in E is defined by the function $\eta_{\tilde{A}}$, called membership function of $\tilde{A}$, from E to the real unit interval [0,1]. $\eta_{\tilde{A}}(x)$ is called the grade or the membership degree of x, $\forall x \in \tilde{A}$ (cf [11]).

Definition 3: Let $\tilde{A}$ fuzzy set on E. The $\alpha-$cut of $\tilde{A}$ denoted $\tilde{A}_\alpha$ the support $\sup(\tilde{A})$, the height $h(\tilde{A})$ and the core $(\tilde{A})$ are crisp sets defined as follows:

$$\tilde{A}_\alpha = \{x \in E : \eta_{\tilde{A}}(x) \geq \alpha\} \tag{4}$$

$$\sup(\tilde{A}) = \{x \in E : \eta_{\tilde{A}}(x) > 0\} \tag{5}$$

$$h(\tilde{A}) = \max\{\eta_{\tilde{A}}(x) : x \in E\} \tag{6}$$

$$\text{core}(\tilde{A}) = \{x \in E : \eta_{\tilde{A}}(x) = 1\} \tag{7}$$

Definition 4: A fuzzy set on a universe is said to be *normal* if:

$$h(\tilde{A}) = 1 \tag{8}$$

that is, $\exists m \in \tilde{A} : \eta_{\tilde{A}}(m) = 1$. In these conditions, m is called *modal value* of the fuzzy set $\tilde{A}$.

Definition 5: A fuzzy set $\tilde{A}$ on the universe $E = \mathbb{R}$ is said to be *convex* iff:

$$\forall x, y \in \tilde{A}, \forall \lambda \in [0,1] : \eta_{\tilde{A}}(\lambda x + (1-\lambda)y) \geq \min\{\eta_{\tilde{A}}(x), \eta_{\tilde{A}}(y)\} \tag{9}$$

3.2 Fuzzy numbers

Definition 6: A fuzzy set $\tilde{A}$ on a universe E is called a fuzzy number if it satisfies the following conditions:

1. $E = \mathbb{R}$

2. $\tilde{A}$ is normal

3. $\tilde{A}$ is convex

4. The membership function $\eta_{\tilde{A}}$ is piecewise continuous

3.3 Fuzzy numbers of type L: R of L: R type

Definition 7: A fuzzy set $\tilde{A}$ is said to be of L–R type if there exists three reals $m, a > 0, b > 0$ and two continuous and decreasing positive functions L and R from $\mathbb{R}$ in [0,1] such that: L(0) = R(0) =1

$$L(1) = 0, \quad \text{or} \quad L(x) > 0, \text{with} \lim_{x \to \infty} L(x) = 0 \tag{10}$$

$$R(1) = 0, \quad \text{or} \quad R(x) > 0, \text{with} \lim_{x \to \infty} R(x) = 0 \tag{11}$$

$$\eta_{\tilde{A}}(x) = \begin{cases} L\left(\frac{m-x}{a}\right) & \text{if}\, x \in [m-a, m] \\ R\left(\frac{m-x}{b}\right) & \text{if}\, x \in [m, m+b] \\ 0 & \text{otherwise} \end{cases} \tag{12}$$

The L–R representation of a fuzzy number $\tilde{A}$ is $\tilde{A} = \langle m, a, b \rangle_{L-R}$, m is called the modal value of $\tilde{A}$. a and b are called respectively the *left spread* and *right spread* of $\tilde{A}$.

By convention, $\langle m,0,0 \rangle_{L-R}$ is the ordinary real number m; called also fuzzy singleton. The support of $\tilde{A}$ is the open interval:
$\sup(\tilde{A}) =]m-a, m] \cup [m, m+b[=]m-a, m+b[$.

From the Definition (8) and the expression (12) of $\eta_{\tilde{A}}$, the support of $\tilde{A}$ is determined by the open following interval:

$$\sup(\tilde{A}) =]m-a, m] \cup [m, m+b[=]m-a, m+b[\tag{13}$$

3.4 Arithmetic of fuzzy numbers of L: R Type

3.4.1 Addition and subtraction of fuzzy numbers of L: R type

According to [10], if there exists two fuzzy numbers of the same L–R type. $\tilde{A} = \langle m, a, b\rangle_{L-R}$ and $\tilde{B} = \langle n, c, d\rangle_{L-R}$; then their sum and their difference are also fuzzy numbers of L–R type given respectively by:

$$\tilde{A} \oplus \tilde{B} = \langle m+n, a+c, b+d\rangle_{L-R} \tag{14}$$

$$\tilde{A} \ominus \tilde{B} = \langle m-n, a+c, b+d\rangle_{L-R} \tag{15}$$

3.4.2 Multiplication and division

According to [12], if there exist two fuzzy numbers of the same L–R type. $\tilde{A} = \langle m, a, b\rangle_{L-R}$ and $\tilde{B} = \langle n, c, d\rangle_{L-R}$; then:

$$\tilde{A} \odot \tilde{B} \approx \langle mn, mc, +na - ac, md + nb + bd\rangle_{L-R} \tag{16}$$

$$\frac{\tilde{A}}{\tilde{B}} = \frac{\langle m, a, b\rangle_{L-R}}{\langle n, c, d\rangle_{L-R}} \approx \left\langle \frac{m}{n}, \frac{md}{n(n+d)} + \frac{a}{n} - \frac{ad}{n(n+d)}, \frac{mc}{n(n-c)} + \frac{b}{n} + \frac{bc}{n(n-c)} \right\rangle_{L-R} \tag{17}$$

The product and the quotient of two numbers of the same type L–R are obtained by the secant approximation of Hanss [14], whose kernel and the support, for the quotient are given by:

$$\ker\left(\frac{\tilde{A}}{\tilde{B}}\right) = \frac{m}{n} \tag{18}$$

$$supp\left(\frac{\tilde{A}}{\tilde{B}}\right) = \left]\frac{m-a}{n+d}, \frac{m+b}{n-c}\right[\tag{19}$$

3.5 Fuzzy triangular numbers

Definition 8: A fuzzy number $\tilde{A}$ is said to be a *fuzzy triangular number* iff there exists three real numbers $a < b < c$ such that:

$$\eta_{\tilde{A}}(x) = \begin{cases} \left(\frac{x-a}{b-a}\right) & \text{if } a \leq x \leq b \\ \left(\frac{c-x}{c-b}\right) & \text{if } b \leq x \leq c \\ 0 & \text{otherwise} \end{cases} \tag{20}$$

Remark 1.

a. Every fuzzy triangular number is noted by: $\tilde{A} = (a, b, c)$ or $\tilde{A} = (a|b|c)$

b. Every fuzzy triangular number of L–R type and the fuzzy L–R representation of $\tilde{A}$ is:

$$\tilde{A} = (a, b, c) = \langle b, b-a, c-b \rangle_{L-R} \quad (21)$$

4. Fuzzy function

In this section, let us define, in general, a fuzzy function of one real variable as follows:

$$\tilde{\psi} : \mathcal{D} \to \mathcal{F}(\mathbb{R}) : x \mapsto \tilde{\psi}(x) \quad (22)$$

where $\mathcal{F}(\mathbb{R})$ the set of fuzzy functions defined on $\mathbb{R}$.

a. *Alpha-cut of* $\tilde{\psi}(x)$

The α-cup representation of $\tilde{\psi}(x)$ is:

$$\tilde{\psi}(x)_\alpha = \tilde{\psi}(x, \alpha) = \left[\tilde{\psi}^L(x, \alpha), \tilde{\psi}^U(x, \alpha)\right], \alpha \in [0, 1] \quad (23)$$

b. *Kernel of* $\tilde{\psi}(x)$

The kernel of $\tilde{\psi}(x)$, also called modal of $\tilde{\psi}(x)$ is defined by:

$$\ker\left(\tilde{\psi}(x)_{\alpha=1}\right) = \tilde{\psi}^L(x, 1) = \tilde{\psi}^U(x, 1) \quad (24)$$

c. *Support of* $\tilde{\psi}(x)$

The support of $\tilde{\psi}(x)$ is defined by:

$$supp\left(\tilde{\psi}(x)_{\alpha=0}\right) = \tilde{\psi}(x, 0) = \left[\tilde{\psi}^L(x, 0), \tilde{\psi}^U(x, 0)\right] \quad (25)$$

d. *Membership function of* $\tilde{\psi}(x)$

The membership function of the fuzzy function $\tilde{\psi}(x)$ is defined by:

$$\eta_{\tilde{\psi}(x)} = \begin{cases} \left(\tilde{\psi}^L\right)^{-1}(x, \varsigma_x) & \text{if } \tilde{\psi}^L(x, 0) \leq \varsigma_x \leq \tilde{\psi}^L(x, 1) \\ \left(\tilde{\psi}^U\right)^{-1}(x, \varsigma_x) & \text{if } \tilde{\psi}^U(x, 1) \leq \varsigma_x \leq \tilde{\psi}^L(x, 0) \\ 0 & \text{otherwise} \end{cases} \quad (26)$$

$$\eta_{\tilde{\psi}(x)} = \begin{cases} \dfrac{\varsigma_x - \tilde{\psi}^L(x,)}{\tilde{\psi}^L(x, 1) - \tilde{\psi}^L(x, 0)} & \text{if } \tilde{\psi}^L(x, 0) \leq \varsigma_x \leq \tilde{\psi}^L(x, 1) \\ \dfrac{\tilde{\psi}^U(x, 0) - \varsigma_x}{\tilde{\psi}^L(x, 0) - \tilde{\psi}^L(x, 1)} & \text{if } \tilde{\psi}^U(x, 1) \leq \varsigma_x \leq \tilde{\psi}^L(x, 0) \\ 0 & \text{otherwise} \end{cases} \quad (27)$$

5. Description of the L: R method and procedure of computing

5.1 Description of L: R method

Let us consider the classical Markovian queue M/M/1 defined in Section (2.2) and assume that the arrival rate λ and service rate μ are triangular fuzzy numbers denoted

$\tilde{\lambda} = (\lambda_1|\lambda_2|\lambda_3)$ and $\tilde{\mu} = (\mu_1|\mu_2|\mu_3)$, respectively. In this case, these rates are imprecise (or fuzzy) and also make the performance measures of the transient fuzzy functions and, we note:

$$\tilde{\psi}(t) = \tilde{f}\left(t, \tilde{\lambda}, \tilde{\mu}\right) \tag{28}$$

Where t is a real variable called time and $\tilde{\lambda}$ and $\tilde{\mu}$ are fuzzy variables. In this case, the queueing model becomes a fuzzy Markovian queue FM/FM/1, where FM is a fuzzy exponential distribution.

To determine the fuzzy performance measure $\tilde{\psi}(t)$, the L–R method proceeds as follows:

5.2 Procedure

Determine the L–R expressions of the fuzzy rates $\tilde{\lambda}$ and $\tilde{\mu}$ and substitute them:

$$\tilde{\psi}(t) = \tilde{f}\left(t, \tilde{\lambda}, \tilde{\mu}\right) \tag{29}$$

Apply the arithmetic of fuzzy numbers of (14)–(17) in (28) and we find:

$$\tilde{\psi}(t) = \langle m(t), \varphi(t), \omega(t)\rangle_{L-R} \tag{30}$$

Where $m(t)$ is the modal function of $\tilde{\psi}(t)$ (or the kernel of $\tilde{\psi}(t)$) and where $\varphi(t)$ and $\omega(t)$ represent respectively the *left spread* and *right spread* of $\tilde{A}$..

The support of $\tilde{\psi}(t)$ is:

$$supp(\tilde{\psi}(t)) =]m(t) - \varphi(t), m(t) + \omega(t)[\tag{31}$$

And its kernel (or modal) is:

$$\ker(\tilde{\psi}(t)) = m(t) \tag{32}$$

6. Numerical example

6.1 Statement

In a referral Hospital, an ophthalmologist doctor consults patients on odd days each week from $10h^{00}$ to $13h^{30}$. The patients arrive there following a Poisson distribution with parameter $\tilde{\lambda}$ and the doctor's consultation following an expo-negative distribution with parameter μ The fuzzy parameters $\tilde{\lambda}$ and μ are such that $\frac{\tilde{\lambda}}{\tilde{\mu}}$ is approximately 0.4. We note that $\tilde{\rho} = \frac{\tilde{\lambda}}{\tilde{\mu}}$ is the fuzzy traffic intensity. We further warn that this traffic intensity is a triangular fuzzy number and is denoted by $\tilde{\rho} = (0.3|0.4|0.5)$.

6.2 Questions

- Determine the following transient performance measures:
- A1. The average number of patients in the system.

- A2. The average time of stay of patients in the system.
- B. Give the graphical representation of these performance measures.

6.3 Solution

A careful reading of our example reveals that it is a fuzzy Markovian queue noted FM/FM/1 with a single server and infinite capacity. The fuzzy traffic intensity being about 0.4 implies that the fuzzy rates $\tilde{\lambda}$ and $\tilde{\mu}$ are about 2 and 5, respectively. By assumption, since the fuzzy traffic intensity $\tilde{\rho}$ is a triangular fuzzy number, the rates $\tilde{\lambda}$ and $\tilde{\mu}$ are also triangular fuzzy numbers and can be written (cf. Remark 1, item a):

$$\tilde{\lambda} = (1|2|3) \quad \text{and} \quad \tilde{\mu} = (4|5|6)$$

Thus, in fuzzy model, the rates $\tilde{\lambda}$ and $\tilde{\mu}$ are fuzzy variables and the performance measures $\tilde{N}_S$ and $\tilde{T}_S$ are fuzzy time functions defined by:

$$\tilde{N}_S(t) = \tilde{f}_1(t, \tilde{\lambda}, \tilde{\mu}) = \frac{\tilde{\lambda}(1 - \exp(-(\tilde{\mu} - \tilde{\lambda})t))}{\tilde{\mu} - \tilde{\lambda}} \tag{33}$$

$$\tilde{T}_S(t) = \tilde{f}_2(t, \tilde{\lambda}, \tilde{\mu}) = \frac{\tilde{\lambda} - \tilde{\lambda}\exp(-(\tilde{\mu} - \tilde{\lambda}))}{\tilde{\mu} - \tilde{\lambda}\left[\tilde{\lambda} + (\tilde{\mu} - \tilde{\lambda})\exp(-(\tilde{\mu} - \tilde{\lambda})t)\right]} \tag{34}$$

To evaluate these performance parameters by L–R method, we proceed as follows:

i. Let us determine the L–R expressions of fuzzy rates:

$\tilde{\lambda} = (1|2|3)$ and $\tilde{\mu} = (4|5|6)$ and we have according to (21):

$$\tilde{\lambda} = \langle 2,1,1 \rangle_{L-R} \quad \text{and} \quad \tilde{\lambda} = \langle 5,1,1 \rangle_{L-R} \tag{35}$$

ii. Let us substitute the expressions of (35) into (33) and (34), and use the operations in (14)–(17) to obtain successively:

$$\begin{aligned}
\tilde{N}_S(t) &= \frac{\tilde{\lambda}(1 - \exp(-(\tilde{\mu} - \tilde{\lambda})t))}{\tilde{\mu} - \tilde{\lambda}} \\
&= \frac{\langle 2,1,1 \rangle_{L-R} - \langle 2,1,1 \rangle_{L-R}\exp(-X)}{\langle 5,1,1 \rangle_{L-R} - \langle 2,1,1 \rangle_{L-R}} \qquad \text{with} \quad X = -\langle 3,2,2 \rangle_{L-R} t \\
&= \frac{\langle 2 - 2\exp(X), 1 + \exp(X), 1 + \exp(X) \rangle_{L-R}}{\langle 3,2,2 \rangle_{L-R}} \\
&\approx \left\langle \frac{2 - 2e^X}{3}, \frac{(2 - 2e^X)X2}{3(3+2)} + \frac{1 + e^X}{3} - \frac{(1 + e^X)X2}{3(3+2)}, \frac{(2 - 2e^X)X2}{3(3+2)} + \frac{1 + e^X}{3} + \frac{(2 - 2e^X)X2}{3(3+2)} \right\rangle_{L-R} \\
&= \left\langle \frac{2 - 2e^X}{3}, \frac{7 - e^X}{15}, \frac{7 - e^X}{3}, \right\rangle_{L-R} \\
&= \langle 0.67 - 0.67e^{-3t}, 0.47 - 0.1e^{-3t}, 2.3 - 0.3e^{-3t} \rangle_{L-R}, \quad X = -3t
\end{aligned}$$

since $X = -\langle 3,2,2\rangle t_{L-R} = -3t$ when we change the L–R writing of X to the α-cut writing, with $\alpha = 0 (t \geq 0)$

$$\tilde{T}_S(t) = \frac{\tilde{\lambda} - \tilde{\lambda}\exp(-(\tilde{\mu} - \tilde{\lambda}))}{\tilde{\mu} - \tilde{\lambda}[\tilde{\lambda} + (\tilde{\mu} - \tilde{\lambda})\exp(-(\tilde{\mu} - \tilde{\lambda})t)]}$$

$$= \frac{\langle 2,1,1\rangle_{L-R} - \langle 2,1,1\rangle_{L-R}\exp(-X)}{\langle 3,2,2\rangle_{L-R}[\langle 2,1,1\rangle_{L-R} - \langle 3,2,2\rangle_{L-R}\exp(-X)]} \quad \text{with} \quad X = -\langle 3,2,2\rangle_{L-R} = -3t$$

$$= \frac{\langle 2 - 2e^X, 1 + e^X, 1 + e^X\rangle_{L-R}}{\langle 3,2,2\rangle_{L-R}[\langle 2 - 3e^X, 1 + 2e^X, 1 + 2e^X\rangle_{L-R} e^X]}$$

$$= \frac{\langle 2 - 2e^X, 1 + e^X, 1 + e^X\rangle_{L-R}}{\langle 6 + 9e^X, 5 + 8e^X, 9 + 16e^X >_{L-R}}$$

$$\approx \left\langle \frac{2 - 2e^X}{6 + 9e^X}, \frac{(2 - 2e^X)(9 + 16e^X)}{(6 + 9)[6 + 9e^X + 9 + 16e^X]} + \frac{1 + e^X}{6 + 9e^X} - \frac{(1 + e^X)(9 + 16e^X)}{(6 + 9e^X)[6 + 9e^X - 5 - 16e^X]}, \right.$$

$$\left. \frac{(2 - 2e^X)(5 + 8e^X)}{(6 + 9e^X)[6 + 9e^X - 5 - 8e^X]} + \frac{1 + e^X}{6 + 9e^X} + \frac{(1 + e^X)(5 + 8e^X)}{(6 + 9e^X)[6 + 9e^X - 5 - 8e^X]} \right\rangle_{L-R}$$

$$= \left\langle \frac{2 - 2e^X}{6 + 9e^X}, \frac{24 + 29e^X - 23e^X}{(6 + 9e^X)(15 + 25e^X)}, \frac{16 + 21e^X - 7e^{2X}}{(6 + 9e^X)(1 + e^X)} \right\rangle_{L-R}$$

with $X = -\langle 3,2,2\rangle_{L-R} t = -3t$ and $t \geq 0$.

6.4 Supports and modals

$$supp(\tilde{N}_S(t)) =](0.67 - 0.67e^X) - (0.47 - 0.1e^X), (0.67 - 0.67e^X) + (2.3 - 0.3e^X)[$$
$$=]0.2 - 0.6e^X, 3 - e^X[, \quad X = -3t, t \geq 0$$

$$supp(\tilde{T}_S(t)) = \left] \frac{2 - 2e^X}{6 + 9e^X} - \frac{24 + 29e^X - 23e^X}{(6 + 9e^X)(15 + 25e^X)}, \frac{2 - 2e^X}{6 + 9e^X} + \frac{16 + 21e^x - 7e^X}{(6 + 9e^X)(1 + e^X)} \right[$$
$$= \left] \frac{1 - 3e^X}{15 + 25e^X}, \frac{3 - e^X}{1 + e^X} \right[, \quad X = -3t, t \geq 0$$

$$\ker(\tilde{N}_S(t)) = 0.67 - 0.67e^X, \quad \text{where} \quad X = -3t, t \geq 0$$

$$\ker(\tilde{T}_S(t)) = \frac{2 - 2e^X}{6 + 9e^X}, \quad X = -3t, t \geq 0$$

6.5 Graphical representations of $\tilde{N}_S(t)$ and $\tilde{T}_S(t)$

The graphical representation of the fuzzy functions $\tilde{N}_S(t)$ and $\tilde{T}_S(t)$ is made from their membership functions when their supports and modes are known. Referring to (27), we obtain:

- $$\eta_{\tilde{N}_{s(t)}}(t, x_t) = \begin{cases} \dfrac{x_t - \tilde{N}_s^L(t,0)}{\tilde{N}_s^L(t,1) - \tilde{N}_s^L(t,0)} & \text{if } \tilde{N}_s^L(t,0) \le x_t \le \tilde{N}_s^L(t,1), t \ge 0 \\ \dfrac{\tilde{N}_s^U(t,0) - x_t}{\tilde{N}_s^U(t,0) - \tilde{N}_s^U(t,1)} & \text{if } \tilde{N}_s^U(t,1) \le x_t \le \tilde{N}_s^U(t,0), t \ge 0 \\ 0 & \text{otherwise} \end{cases}$$

$$= \begin{cases} \dfrac{x_t + 0.6e^{-3t} - 0.2}{0.5 - 0.1e^{-3t}} & \text{if } 0.2 - 0.6e^{-3t} \le x_t \le 0.67 - 0.67e^{-3t}, t \ge 0 \\ \dfrac{(3 - e^{-3t}) - x_t}{2.3 - 0.3e^{3t}} & \text{if } 0.67 - 0.67e^{-3t} \le x_t \le 3 - e^{-3t}, t \geq 0 \\ 0 & \text{otherwise} \end{cases}$$

- $$\eta_{\tilde{T}_{s(t)}}(t, x_t) = \begin{cases} \dfrac{x_t - \tilde{T}_s^L(t,0)}{\tilde{T}_s^L(t,1) - \tilde{T}_s^L(t,0)} & \text{if } \tilde{T}_s^L(t,0) \le x_t \le \tilde{T}_s^L(t,1), t \ge 0 \\ \dfrac{\tilde{T}_s^U(t,0) - x_t}{\tilde{T}_s^U(t,0) - \tilde{T}_s^U(t,1)} & \text{if } \tilde{T}_s^U(t,1) \le x_t \le \tilde{T}_s^U(t,0), t \ge 0 \\ 0 & \text{otherwise} \end{cases}$$

$$= \begin{cases} \left(\dfrac{90 + 285e^{-3t} + 225e^{-6t}}{24 + 29e^{-3t} - 23e^{-6t}}\right)x_t - \dfrac{6 - 9e^{-3t} - 27e^{-6t}}{24 + 29e^{-3t} - 23e^{-6t}} & \text{if } \dfrac{1 - 3e^{-3t}}{15 + 25e^{-3t}} \le x_t \le \dfrac{2 - 2e^{-3t}}{6 + 9e^{-3t}}, t \ge 0 \\ -\left(\dfrac{6 + 15e^{-3t} + 9e^{-6t}}{24 + 21e^{-3t} - 23e^{-6t}}\right)x_t + \dfrac{18 + 21e^{-3t} - 9e^{-6t}}{16 + 21e^{-3t} - 7e^{-6t}} & \text{if } \dfrac{2 - 2e^{-3t}}{6 + 9e^{-3t}} \le x_t \le \dfrac{3 - e^{-3t}}{1 + e^{-3t}}, t \ge 0 \\ 0 & \text{otherwise} \end{cases}$$

6.5.1 Graphics

6.5.2 Interpretation of results

Figure 1, shows that the support of the function $\tilde{N}_s(t)$ is an open interval from $\tilde{N}_s^L(t,0) = 0.2 - 0.6e^{-3t}$ to $\tilde{N}_s^U(t,0) = 3 - e^{-3t}$, $(t \ge 0)$ this means that the function average number of patients in the system is a fuzzy function.

It is impossible for the curve of the fuzzy function $\tilde{N}_s(t)$ to be below the curve of $\tilde{N}_s^L(t,0)$ nor above that of $\tilde{N}_s^U(t,0)$. The curve of its modal function $\tilde{N}_s^U(t,0) = 0.67 - 0.67e^{-3t}(t \ge 0)$, is the most likely curve for the function $\tilde{N}_s(t)$.

Similarly, **Figure 2**, indicates that the curve for the fuzzy function $\tilde{T}_s^L(t)$, the average time of patients' stay in the system, is approximately between the curves of equations $\tilde{T}_s^L(t,0) = \frac{1-3e^{-3t}}{15+25e^{-3t}}$ and $\tilde{T}_s^U(t,0) = \frac{3-3e^{-3t}}{1+e^{-3t}}$ $(t \ge 0)$. In other words, the curve of $\tilde{T}_s(t)$ cannot go below that of $\tilde{T}_s^L(t,0)$ nor can it go above that of $\tilde{T}_s^U(t,0)$.

In the classical, we can say that the average waiting time of patients in the system is approximately between 0.1 hours ($\simeq$ 6 min) and 3 hours ($\simeq$ 180 min).

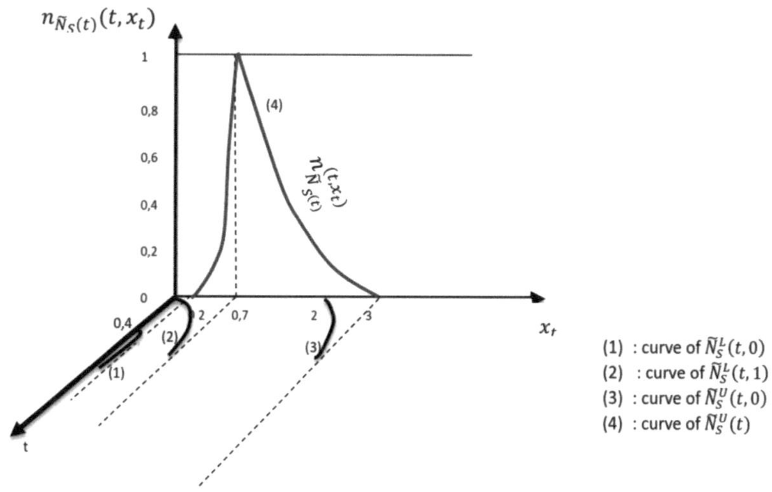

Figure 1.
Membership function of the fuzzy function $\tilde{N}_s(t)$ average number of patients in the system at a time $t(t \geq 0)$.

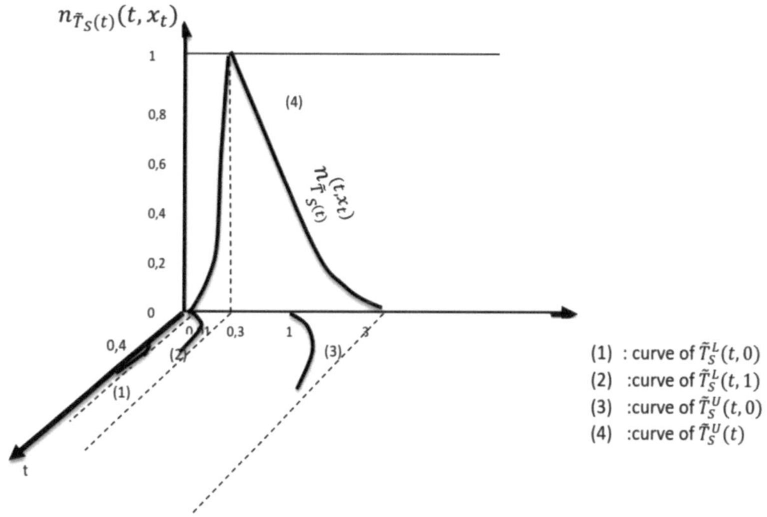

Figure 2.
Membership function of the fuzzy function $\tilde{T}_s(t)$ average time of patients' stay in the system at a time $t(t \geq 0)$.

This means in other words that the average patient waiting time in the system cannot go below 6 min nor above 180 min. The most likely mean patient waiting time in the system is 0.3 hours ($\simeq$ 18 min).

7. Conclusion

At the end of this chapter related to "the calculation of the performance parameters of the fuzzy Markovian queueing system FM/FM/1 in transient state by the L–R method," it was a question of evaluating these performance parameters of the queueing system considered in transient regime by means of a method called the L–R method based primarily on the arithmetic of the fuzzy numbers of type L–R restricted to the secant approximations.

The L–R method facilitated us to find, in transient state, the L–R representation of the performance parameters, their support and mode as well as their membership functions which allowed us to represent graphically the performance parameters of the queueing model in study in the space in three dimensions. This is the originality of this scientific work.

In this chapter, a fuzzy queue (or waiting system) analysis method called L–R method, essentially based on the fuzzy L–R arithmetic deduced from secant approximations has been studied. The L–R representation was used to find the performance measures of the studied model. Using this method, the fuzzy functions were computed for the fuzzy expectation model FM/FM/1 and the results are found in L–R representation. Under this representation, the fuzzy results give much more reliable information than the relaxed alpha-cuts method which will be the subject of a future research study. The solutions obtained show that this method has three major advantages: it is short, convenient and flexible compared to other methods used in this field.

We are sure that the L–R method can still help, to obtain results of other similar problems posed in this field in the framework of evaluating the performance parameters of fuzzy Markovian expectation models in transient regime.

An illustrative example was given in the medical field to show the relevance of this study in operational research and in particular in queueing.

Author details

Rostin Mabela Makengo Matendo[1*], Jean Alonge W'Omatete[2], Herman Matondo Mananga[1], Jean Pierre Mukeba Kanyinda[3] and Baudouin Adia Leti Mawa[4]

1 Department of Mathematics and Computer Science, University of Kinshasa, DR Congo

2 Department of Mathematics and Physics, Higher Institute of Education of Wembo-Nyama (ISP Wembo-Nyama), DR Congo

3 Department of Mathematics and Computer Science, Higher Institute of Education of Mbuji-Mayi, Mbuji-Mayi (ISP Mbuji-Mayi), DR Congo

4 Higher Institute of Education and Technology of Kinshasa, Kinshasa (ISPT Kinshasa), DR Congo

*Address all correspondence to: rostin.mabela@unikin.ac.cd

References

[1] Baynat B. Théorie des files d'attente: des chaaînes de Markov aux réseaux à forme produit. Réseaux et télécommunications. Paris: Hermes Science Publications; 2000

[2] Ning Y, Zhao R. Analysis on random fuzzy queueing systems with finite capacity. In: The Ninth International Conference on Electronic Business, Macau, November 30 – December 4, 2009. pp. 1-7

[3] Ritha W, Robert L. Application of fuzzy set theory to retrial queues. International Journal of Algorithms Computing and Mathematics. 2009; **2**(4):11

[4] Ritha W, Menon BS. Fuzzy n policy queues with infinite capacity. 2011

[5] Li R-J, Lee ES. Analysis of fuzzy queues. Computers & Mathematics with Applications. 1989;**17**(7):1143-1147

[6] Li RJ, Lee ES. Proceedings nafips. 1998. pp. 158–162

[7] Kao C, Li C-C, Chen S-P. Parametric programming to the analysis of fuzzy queues. Fuzzy Sets and Systems. 1999; **107**(1):93-100

[8] Palpandi B, Geetharamani G. Evaluation of Performance Measures of Bulk Arrival Queue with Fuzzy Parameters Using Robust Ranking Technique. International Journal of Computational Engineering Research, Vol. 03(10), India, 2013. pp. 55-57

[9] Wang T-Y, Yang D-Y, Li M-J. Fuzzy analysis for the tv-policy queues with infinite capacity. International Journal of Information and Management Sciences. 2010;**21**:03

[10] Dubois D, Prade HM. Fuzzy Sets and Systems: Theory and Applications. Vol. 144. Amsterdam: Elsevier Science; 1980

[11] Mukeba Kanyinda JP, Matendo RM, Lukata BUE, Ibula DN. Fuzzy eigenvalues and fuzzy eigenvectors of fuzzy markov chain transition matrix under max-min composition. Journal of Fuzzy Set Valued Analysis. 2015;**2015**: 25-35

[12] Mabela R, Ulungu B, Mukeba JP. Sur la multiplication des nombres flous de type l-r. Annals of Faculty of Sciences. 2014;**2014**(1):33-40

[13] Boucherie RJ, van Dijk NM. Queueing networks: A fundamental approach. In: International Series in Operations Research & Management Science. US: Springer; 2010

[14] Hanss M. Applied Fuzzy Arithmetic: An Introduction with Engineering Applications. Berlin Heidelberg: Springer; 2005

Chapter 4

Development of L-Group Theory

Iffat Jahan

Abstract

In this work, we present a systematic and successful development of L-group theory. A universal construction of a generated L-subgroup has been provided by using level subsets of given L-subsets. This construction allows us to define and study commutator L-subgroups, normalizer of an L-subgroup, nilpotent L-subgroups, solvable L-subgroups, normal closure of an L-subgroup. All these concepts and their inter-relationships have been presented. Here we mention that in this work we also exhibit a characterization of solvable L-subgroup with the help of a series of L-subgroups such that at each level, the factor groups of level subgroups of their consecutive members are Abelian. This allows us to introduce the notion of a supersolvable L-subgroup by using the factors of level subgroups at each level of a subinvariant series of an L-subgroup. Also, by using successive normal closures, we transfinitely define a series called the normal closure series of the L-subgroup. It has been shown that it is the fastest descending normal series containing given L-subgroup. This sets the ground for the development of subnormality in L-group theory. In the last, we study the notion of subnormal L-subgroups.

Keywords: L-Subgroup, Generated L-subgroups, Commutator L-subgroup, Characteristic L-subgroup, Nilpotent L-subgroup, Solvable L-subgroup, supersolvable L-subgroup, Subnormal L-subgroup,

1. Introduction

After Rosenfeld [1] introduced his notion of fuzzy subgroups of an ordinary group, there was a great activity in the investigations of fuzzy algebraic structures. Various aspects of fuzzy algebra were explored. Specially in the field of fuzzy group theory, several researchers contributed towards development [2–20]. However, the researchers were not coherent and their studies suffered from one kind of incompatibility or the other. In the year 1981, Liu [21] introduced his notion of normality of a fuzzy subgroup in an ordinary group. Soon after that Wu [22] came up with an idea and just gave a hint of pursuing studies of fuzzy subgroups of a fuzzy group by providing the definition of a fuzzy normal subgroup in a fuzzy group. Afterwards, this idea was further taken up only by Martinez [23, 24]. Following this approach, the theory of L-subgroups is developed in a systematic and consistent manner in our papers [10, 25–34].

In fact, L-group theory came into existence as an answer to the problems posed by fuzzy group theory and the theories of other fuzzy algebraic structures. The development of fuzzy group theory was hampered mainly because of two reasons. Firstly, it was due to some inherent problems that the analogs of some of the concepts and results of classical group theory were not even formulated in this theory (for details, see Section 7, Discussion and Analysis). Moreover, various types of series such as derived series, descending central series, normal closure series etc. could not be formulated. Secondly, since most of the concepts studied in fuzzy algebraic structures were generically defined, Tom Head [35] proposed his well known metatheorem (which is based on the concept of Rep function) to extend the results of classical algebra to fuzzy setting. Therefore, the results of fuzzy group theory became simple instances of an application of metatheorem. Throughout the development of L-group theory, the above two drawbacks of fuzzy group theory have been very well taken care of (for details, see Section 7, Discussion and Analysis) and we are in a position to put forward a theory parallel to classical group theory. Thus a consistent theory came into existence. The subject matter discussed in this chapter, in particular; the join problem of L-subgroups provides sufficient testimony to its success [10].

As an application and motivation, here we mention that if we replace the lattice L, in our work by the closed unit interval $[0, 1]$, *then we retrieve the corresponding version for fuzzy group theory. Also, as an application of this theory we mention that if we replace the lattice L by the two elements set* $\{0, 1\}$, *then the results of classical group theory follow as simple corollaries of the corresponding results of L-group theory. Moreover, this development of L-group theory is beyond the purview of metatheorem, contrary to the development of fuzzy group theory.*

Section 2 provides a list of all the basic definitions and results regarding L-subsets and L-algebraic substructures which are required for the development of the subsequent sections. For the sake of completeness, few definitions from lattice theory have also been incorporated. In papers [10, 25–34], various concepts of L-group theory have been explored.

Section 3 introduces the concept of a normalizer of an L-subgroup of an L-group. In Subsection 3.1, this notion of normalizer is very carefully formulated by using the concept of a coset by an L-point of an L-group. The normalizer of an L-subgroup in an L-group, formulated in this work, is an L-subgroup of the given L-group. This concept of classical group theory was left untouched during the evolution of fuzzy group theory. Although, Mukherjee and Bhattachrya [36, 37] tried to introduce a notion of normalizer of a fuzzy subgroup of an ordinary group, but it turned out to be a crisp subgroup of the given ordinary group. This idea was followed by several researchers [16, 38] in the past, but they were unable to obtain the results, presented in this work, due the fact that they carried out their researches within the framework where the parent structure was an ordinary group. We have shown in our work that for a normal L-subgroup of an L-group, its left and right L-cosets are identical. Also, it has been proved that each L-subset of an L-group commutes with every normal L-subgroup of the given L-group. In the end of this subsection, we state certain properties and the nature of this normalizer under the action of a group homomorphism. In Subsection 3.2, a universal construction of a generated L-subgroup by an L-set has been provided and its relationship with level subsets is investigated. This construction along with the construction of commutator L-subsets, studied in Section 4, allows us to define a commutator L-subgroup. In Subsection 3.3, again by replacing the parent structure of an ordinary group by an L-subgroup, we formulate the concept of a characteristic L-subgroup of an L-group. After obtaining the level subset characterizations of this concept, we establish some group theoretic analogs. Then, we construct various types of lattices and sublattices of characteristic L-subgroups. Finally, in this section we are able to establish that a characteristic L-subgroup of a normal L-subgroup is normal. Also, the well known property of transitivity of characteristic subgroup of classical group theory is extended to the L-setting. However, the same could not be even formulated in the works of earlier researchers [14, 39–41].

DOI: http://dx.doi.org/10.5772/intechopen.110387

Section 4 starts with the notion of a commutator L-subset and commutator L-subgroup of an L-group. It is worthwhile to mention here that, earlier Gupta and Sarma [8] extended the notion of commutator subgroups in fuzzy setting which was utilized to formulate the concepts of descending central chain and derived chain in their further studies [9, 10]. However, the above mentioned studies have been carried out within the framework where the underlying group is an ordinary group. Therefore, they lost certain compatibility with other fuzzy algebraic notions. Here, we obtain some group theoretic analogs of commutator subgroups, we state a property of infimums of the set product of two L-subsets which is used in the further development of the subject matter. The whole development is justified by the level subset and strong level subset characterizations of commutator L-subgroups. Further in this section, we introduce the concept of a descending central chain of an L-subgroup by making the use of the notion of commutators. Then this, in turn, is used to define the notion of nilpotent L-subgroups of an L-group. Here the concept of the trivial L-subgroup of an L-subgroup comes into play. The members of the descending central chain are normal L-subgroups in their preceding ones in the sense of Wu [22]. Then, we present some peculiarities of L-setting which will be discussed in the end of this chapter. The level subset and strong level subset characterizations of these notions justify these extensions. The concept of the central chain of an L-subgroup is also introduced with the help of the trivial L-subgroup of the given L-group which is followed by analogs of some well known results of classical group theory. In the end of this section, we establish a necessary and sufficient condition for the set product of two trivial L-subgroups to be a trivial L-subgroup (see Theorem 1.81). Finally, this result has been used very effectively to establish a sufficient condition for the set product of two nilpotent L-subgroups to be nilpotent [31]. It has also been shown that the notion of a normalizer of an L-subgroup, which has been introduced in Section 3, is compatible with the notion of nilpotent L-subgroups. That is, nilpotent L-subgroup satisfies normalizer condition [34]. On the other hand, Kim [13] also defined an ascending series of crisp subgroups of an ordinary group to introduce his concept of nilpotent fuzzy subgroups. However, this could not lead to any substantial progress.

Section 5 deals with solvability and supersolvability of L-subgroups. Some more researchers [38, 42, 43] also discussed the notion solvability in the fuzzy setting. In the studies carried out by these authors the concept of normality introduced by Liu [21] is used. Consequently, the parent structure in their studies is an ordinary group, not a fuzzy group. For this purpose we introduce the concepts of derived series and solvable series with the help of the notion of commutator L-subgroups. This is possible because we use the normality in the sense of Wu [22] rather than Liu [21]. We discuss some results pertaining to the members of the derived central chain which are peculiarities of L-setting (Theorem 1.84, Theorem 1.85). All these concepts are justified by their level subset and strong level subset characterizations. Moreover, solvability is also characterized in terms of solvable series and some group theoretic analogs are obtained. Finally, the concept of central series is used to establish the inter connection of nilpotency and solvability of L-subgroups. We also discuss the behavior of homomorphic and inverse homomorphic images of solvable L-subgroups. Next, we define normal and subinvariant L-subgroups of an L-subgroup with Abelian factors. In case when the lattice L is a dense chain, we characterize solvability of an L-subgroup with the help of a normal series with Abelian factors or subinvariant series with Abelian factors. This characterization motivated us to introduce the notion of a supersolvable L-subgroup by using the factors of level subgroups at each level of a subinvariant series of an L-subgroup. Also, commutator L-subgroup of a supersolvable L-subgroup is shown to be nilpotent. In the last, we extend Zassenhaus Theorem to L-setting and utilize it to establish a version of Schreier Refinement Theorem.

Section 6 evolves the concept of conjugacy in L-group theory. Firstly, the conjugate of an L-subgroup by an L-subgroup has been defined. The normal closure of an L-subgroup is defined as the L-subgroup generated by its conjugate by the whole parent L-group. Then, by using successive normal closures, we transfinitely define a series called the normal closure series of the given L-subgroup. Earlier an attempt has been made in [5] to define normal closure of a fuzzy group in an ordinary group. In fact, this is the fuzzy subgroup generated by the union of all the conjugates of the given fuzzy subgroup by crisp points. The concept of this conjugacy comes from [36]. But this idea was not found suitable enough to formulate the successive normal closures and hence it was not found suitable to be applied in the development of subnormal fuzzy subgroups introduced by the same author in [44]. Here, it has been shown that the normal closure series, defined in this work, is the fastest descending normal series containing given L-subgroup. This sets the ground for the development of subnormality in L-group theory [10, 33]. During the course of the development of subnormality, it has also been proved that every L-subgroup of a nilpotent L-subgroup is a subnormal L-subgroup. Finally in order to show the reach of L-group theory, we tackle the well known join problem of subnormal subgroups in L-setting and solve it to the same degree of success as that of classical group theory.

2. Preliminaries

Throughout our work, the system $\langle L, \leq, \wedge, \vee \rangle$ denotes a complete and completely distributive lattice where' $\leq$ ' denotes the partial ordering of L, the join(sup) and the meet(inf) of the elements of L are denoted by '$\vee$' and '$\wedge$', respectively. We shall denote the maximal and the minimal elements of L by 1 and 0 respectively. Moreover, I denotes a non-empty indexing set.

The definition of a completely distributive lattice is well known in literature [45]. Let $\{J_i : i \in I\}$ be any family of subsets of a complete lattice L and F denotes the set of choice functions for J_i, i.e., functions $f : I \to \prod_{i \in I} J_i$ such that $f(i) \in J_i$ for each $i \in I$. Then, we say that L is a completely distributive lattice, if

$$\bigwedge \left\{ \bigvee_{i \in I} J_i \right\} = \bigvee_{f \in F} \left\{ \bigwedge_{i \in I} f(i) \right\}. \tag{1}$$

The above law is known as the completely distributive law. Moreover, a lattice L is said to be infinitely meet distributive if for every subset $\{a_i : i \in I\}$ of L we have

$$a \bigwedge \left\{ \bigvee_{i \in I} a_i \right\} = \bigvee_{i \in I} \left\{ a \bigwedge a_i \right\}, \tag{2}$$

provided L is join complete. The above law is known as infinitely meet distributive law. The definition of infinitely join distributive lattice is dual of the above definition, i.e., a lattice L is said to be infinitely join distributive if for every subset $\{a_i : i \in I\}$ of L we have

$$a \bigvee \left\{ \bigwedge_{i \in I} a_i \right\} = \bigwedge_{i \in I} \{ a \bigvee a_i \}, \tag{3}$$

provided L is meet complete. The above law is known as infinitely join distributive law. Both the above laws follow from the definition of a completely distributive lattice. The dual of a completely distributive law is valid in a completely distributive lattice whereas the infinitely meet and join distributive laws are independent of each other. In this section, we recall some definitions and results which will be used in the sequel. An L-subset of X is a function from X into L. The set of all L-subsets of X is called the L-power set of X and is denoted by L^X.

For an ordinary subset A of X, its characteristic function defined by:

$$1_A(x) = \begin{cases} 1, & \text{if } x \in A, \\ 0, & \text{if } x \notin A; \end{cases} \tag{4}$$

is an L-subset of X representing A.

Let $\mu \in L^X$. The set $\{\mu(x) : x \in X\}$ is called the image of μ and is denoted by $\mu(X)$ or $Im\mu$. Let $\sup_{x \in X} \mu(x) = a_0$ and $\inf_{x \in X} \mu(x) = t_0$. We call a_0 to be the tip of μ and t_0 to be the tail of μ. We denote the tip and tail of μ by $\sup\mu$ and $\inf \mu$ respectively. Let $a \in L$. Then, $\mu_a = \{x \in X : \mu(x) \geq a\}$ is called a-level set or a-cut of μ. Moreover, $\mu_a^> = \{x \in X : \mu(x) > a\}$ is called a-strong level set (or a-strong cut) of μ. Note that $\mu_a = \phi$, if $a > a_0$ and $\mu_a^> = \phi$, if $a \geq a_0$. Moreover, $\mu_a = X$ if $a \leq t_0$ and $\mu_a^> = X$ if $a < t_0$. Let $Y \subseteq X$. Then, we define $a_Y \in L^X$ as follows:

$$a_Y(x) = \begin{cases} a, & \text{if } x \in Y, \\ 0, & \text{if } x \in X \backslash Y. \end{cases} \tag{5}$$

In particular, if Y is singleton say $\{y\}$, then $a_{\{y\}}$ is called L-point or L-singleton and is denoted by a_y. We say that the L-point $a_y \in \mu$ if $\mu(x) \geq a$. The union $\underset{i \in I}{\cup} \mu_i$ and the intersection $\underset{i \in I}{\cap} \mu_i$ of any family $\{\mu_i : i \in I\}$ of L-subsets of X are, respectively, defined by:

$$\left(\bigcup_{i \in I} \mu_i\right)(x) = \bigvee_{i \in I} \mu_i(x) \text{ and} \left(\bigcap_{i \in I} \mu_i\right)(x) = \bigwedge_{i \in I} \mu_i(x), \tag{6}$$

for each $x \in X$. Let $\eta, \mu \in L^X$. Then, η is said to be contained in μ, if we have $\eta(x) \leq \mu(x)$ for each $x \in X$ and is written as $\eta \subseteq \mu$ or $\mu \supseteq \eta$.

Theorem 1.1 Let $\eta, \theta \in L^X$. Then

i. $\eta \subseteq \theta$ if and only if $\eta_a \subseteq \theta_a$ for each $a \in L$,

ii. $\eta \subseteq \theta$ if and only if $\eta_a^> \subseteq \theta_a^>$ for each $\in L \sim \{1\}$, provided L is a chain.

Theorem 1.2 Let $\{\mu_i\}_{i \in I} \subseteq L^\mu$. Then

i. $\left(\bigcap_{i \in I} \mu_i\right)_a = \bigcap_{i \in I} (\mu_i)_a$ for all $\in L$,

ii. $\left(\bigcup_{i \in I} \mu_i\right)_a \subseteq \bigcup_{i \in I} (\mu_i)_a$ for all $\in L$,

iii. $\left(\bigcup_{i\in I}\mu_i\right)_a^{>} = \bigcup_{i\in I}(\mu_i)_a^{>}$ for all $a\in L\sim\{1\}$, provided L is a chain;

iv. $\left(\bigcap_{i\in I}\mu_i\right)_a^{>} \subseteq \bigcap_{i\in I}(\mu_i)_a^{>}$ for all $a\in L\sim\{1\}$, provided L is a chain.

Let f be a function from X into Y, and let $\mu\in L^X$ and $\nu\in L^Y$. Then, the image $f(\mu)$ of μ under f and the pre-image $f^{-1}(\eta)$ of η under f are L-subsets of Y and X respectively defined by:

$$f(\mu)(y) = \bigvee_{x\in f^{-1}(y)}\{\mu(x) : x\in X\} \text{ and } f^{-1}(\eta)(x) = \eta(f(x)). \tag{7}$$

If $f^{-1}(y) = \phi$, then $f(\mu)(y) = 0$ since the least upper bound of the empty set in L is 0. The set product $\mu\circ\eta$ of $\mu,\eta\in L^S$, where S is a groupoid, is an L-subset of S defined by

$$\mu\circ\eta(x) = \sup_{x=yz}\{\mu(y)\wedge\eta(z)\}. \tag{8}$$

If x cannot be factored as $x = yz$ in S, then $\mu\circ\eta(x)$ being the least upper bound of the empty set in L is 0.

The set L^X of L-subsets of X, together with the operations of union and intersection, is a complete lattice with the partial ordering of L-set inclusion $\subseteq$. Its maximal and minimal elements are 1_X and 0_X, respectively. Here 1_X and 0_X are L-subsets of X which map each element of X to 1 and 0, respectively. Moreover, the lattice $P(X)$ of all subsets of X can be isomorphically embedded into the lattice L^X.

From now onwards, G will denote an arbitrary group with the identity element e. We recall the definitions of an L-subgroup and a normal L-subgroup of the group G.

Definition 1.1 Let $\mu\in L^G$. Then, μ is called an L-subgroup of G if.

i. $\mu(xy)\geq\mu(x)\wedge\mu(y)$ for each $x,y\in G$,

ii. $\mu(x^{-1}) = \mu(x)$ for each $x\in G$.

The set of all L-subgroups of G is denoted by $L(G)$. From the definition, it is clear that the tip of μ is attained at the identity element of G.

Definition 1.2 Let $\mu\in L(G)$. Then, μ is said to to be normal L-subgroup of G if $\mu(xy) = \mu(yx)$ for each $x,y\in G$.

It is well known that the intersection of an arbitrary family of L-subgroups of a group is an L-subgroup of the given group. Hence we have the following definition:

Definition 1.3 Let $\mu\in L^G$. Then, the L-subgroup of G generated by μ, denoted by $\langle\mu\rangle$, is defined as the smallest L-subgroup of G which contains μ, i.e.,

$$\langle\mu\rangle = \cap\{\nu : \mu\subseteq\nu, \nu\in L(G)\}. \tag{9}$$

The set $L(G)$ is a complete lattice under the ordering of L-set inclusion where the meet'$\wedge$' and join'$\vee$' of an arbitrary family $\{\eta_i\}_{i\in I}$ in $L(G)$ are defined, respectively, by:

$$\underset{i\in I}{\wedge}\eta_i = \left(\bigcap_{i\in I}\eta_i\right) \text{ and } \bigvee_{i\in I}\eta_i = \left\langle\bigcup_{i\in I}\eta_i\right\rangle. \tag{10}$$

Let $\eta, \mu \in L^G$ such that $\eta \subseteq \mu$. Then, η is said to be an L-subset of μ. The set of all L-subsets of μ is denoted by L^μ. Moreover, if $\eta, \mu \in L(G)$ such that $\eta \subseteq \mu$, then η is said to be an L-subgroup of μ. The set of all L-subgroups of μ is denoted by $L(\mu)$. Here we mention that the set $L(\mu)$ of all L-subgroups of μ is a complete sublattice of the lattice $L(1_G)$.

Next, we provide level subset and strong level subset characterizations of an L-subgroup of an L-group.

Theorem 1.3 Let $\eta \in L^\mu$. Then,

i. $\eta \in L(\mu)$ if and only if each non-empty level subset η_a is a subgroup of μ_a,

ii. $\eta \in L(\mu)$ if and only if each non-empty strong level subset $\eta_a^>$ is a subgroup of $\mu_a^>$ provided L is a chain.

The following results discuss homomorphic image and pre-image of an L-subgroup:

Theorem 1.4 Let $\eta \in L(\mu)$ and $\mathrm{f} : G \to H$ be a group homomorphism. Then, $\mathrm{f}(\eta)$ is an L-subgroup of $\mathrm{f}(\mu)$.

Theorem 1.5 Let $\eta, \mu \in L(H)$ with $\eta \in L(\mu)$ and $\mathrm{f} : G \to H$ be a group homomorphism. Then, $\mathrm{f}^{-1}(\eta)$ is an L-subgroup of $\mathrm{f}^{-1}(\mu)$.

Let $\eta \in L(\mu)$ be such that η is non-constant and $\eta \neq \mu$. Then, η is said to be a proper L-subgroup of μ. Clearly, η is a proper L-subgroup of μ if and only if η has distinct tip and tail and $\eta \neq \mu$.

Definition 1.4 Let $\eta \in L(\mu)$. Then, η is said to be a trivial L-subgroup of μ if its chain of level subgroups contains only $\{e\}$ and G..

Definition 1.5 Let $\eta \in L(\mu)$. Then, define an L-subset $\eta_{t_0}^{a_0}$ of μ as follows:

$$\eta_{t_0}^{a_0}(y) = \begin{cases} a_0, \text{if } y = e, \\ t_0, \text{if } y \neq e; \end{cases} \tag{11}$$

where $a_0 = \eta(e)$ and $t_0 = \inf \eta$. Clearly, $\eta_{t_0}^{a_0}$ is a trivial L-subgroup of μ and is called the trivial L-subgroup of η.

From now onwards, we denote μ as an L-subgroup of G and where there is no likelihood of any confusion, we shall not mention the underlying group G. We shall call the parent L-subgroup μ to be simply an L-group.

Now, we study the notion of normal L-subgroups of an L-group and its related properties. In the course of development, we obtain the analogs of certain results of classical group theory for normal L-subgroups of an L-group.

Definition 1.6 Let $\eta \in L^\mu$. Then, η is said to be a normal L-subset of μ if

$$\eta(y^{-1}xy) \geq \eta(x) \wedge \mu(y) \text{ for each } x, y \in G. \tag{12}$$

The set of all normal L-subsets of μ is denoted by NL^μ. Moreover, if $\eta \in L(\mu)$, then η is said to be a normal L-subgroup of μ. The set of all normal L-subgroups of μ is denoted by $\mathrm{NL}(\mu)$. The following result follows immediately by the definition of normal L-subsets:

Theorem 1.6 Let $\{\eta_i\}_{i\in I} \subseteq NL^\mu$ be any family. Then,

i. $\bigcap_{i \in I} \eta_i \in NL^\mu$,

ii. $\bigcup_{i \in I} \eta_i \in NL^\mu$.

Next, we provide level subset and strong level subset characterizations of normal L-subsets and normal L-subgroups of an L-group.

Theorem 1.7 Let $\eta \in L^\mu$. Then,

i. $\eta \in NL^\mu$ if and only if each non-empty level subset η_a is a normal subset of μ_a,

ii. $\eta \in NL^\mu$ if and only if each non-empty strong level subset $\eta_a^>$ is a normal subset of $\mu_a^>$ provided L is a chain.

Theorem 1.8 Let $\eta \in L(\mu)$. Then,

i. $\eta \in NL(\mu)$ if and only if each non-empty strong level subset η_a is a normal subgroup of μ_a,

ii. $\eta \in NL(\mu)$ if and only if each non-empty strong level subset $\eta_a^>$ is a normal subgroup of $\mu_a^>$ provided L is a chain.

The following results deal with the homomorphic image and pre-image of a normal L-subgroups:

Theorem 1.9 Let $\eta, \mu \in L(G)$ such that $\eta \in \mathrm{NL}(\mu)$ and $\mathrm{f} : G \to H$ be an onto group homomorphism. Then, $\mathrm{f}(\eta)$ is a normal L-subgroup of $\mathrm{f}(\mu)$.

Theorem 1.10 Let H be a group and $\mu, \eta \in L(H)$ be such that $\eta \in \mathrm{NL}(\mu)$. Let $\mathrm{f} : G \to H$ be a group homomorphism. Then, $\mathrm{f}^{-1}(\eta)$ is a normal L-subgroup of $\mathrm{f}^{-1}(\mu)$.

The notion of sup-property was introduced by A. Rosenfeld [1] in order to extend certain results of classical group theory to fuzzy setting. Thereafter, this technique was employed by researchers in various fields of fuzzy algebraic substructures [46].

Definition 1.7 Let $\eta \in L^\mu$. Then, η is said to have sup-property if for each non-empty subset A of G, there exists $a_0 \in A$ such that $\vee_{a \in A} \eta(a) = \eta(a_0)$. The set of all L-subsets of μ possessing sup-property is denoted by L_s^μ.

3. Generated L-subgroup, normalizer and characteristic L-subgroup of an L-group

The significance of the notions of normalizers, generated subgroups and characteristic subgroups can be found in any standard text in the classical group theory. In this section, we study these concepts within the framework of L-setting.

3.1 Normalizer of an L-group

One of the key notions 'normalizer of a subgroup' of classical group theory was left untouched during the evolution of fuzzy group theory. The notion of normalizer of an L-subgroup of an L-group has been introduced in [26] which, in essence, is comparable with its classical counterpart. This subsection commences with the definition of a coset of an L-subgroup by an L-point.

Definition 1.8 Let $\eta \in L(\mu)$. Then for $a_x \in \mu$, left (right) coset of η in μ is defined as the set product $a_x \circ \eta$ $(\eta \circ a_x)$.

The following is immediate:

Let $\eta \in L(\mu)$. Then, for each $a_x \in \mu$ and $z \in G$

$$a_x \circ \eta(z) = a \wedge \eta(x^{-1}z), \quad \eta \circ a_x(z) = a \wedge \eta(zx^{-1}); \tag{13}$$

i.e. $a_x \circ \eta = \eta \circ a_x$ for each $a_x \in \eta$.

The characterization of an L-subgroup in terms of L-points is obtained in a way similar to classical group theory.

Theorem 1.11 Let $\eta \in L^{\mu}$. Then,

$$\eta \in \mathrm{L}(\mu) \text{ if and only if } a_x \circ b_{y^{-1}} \in \eta \text{ for each } a_x, b_y \in \eta. \tag{14}$$

We can characterize the normality of an L-subgroup of a given L-group in terms of these 'cosets' as follows. We observe here that in case of normal L-subgroup, the left coset and the right coset by an L-point are identical.

Theorem 1.12 Let $\eta \in L(\mu)$. Then,

$$\eta \in NL(\mu) \text{ if and only if } a_x \circ \eta = \eta \circ a_x \text{ for each } L\text{-point } a_x \in \mu. \tag{15}$$

Clearly for $\eta \in L(\mu)$, $\eta \in NL(\eta)$. Also, we observe that for $a_x, b_y \in \mu$,

$$a_x \circ b_y = (a \wedge b)_{xy}. \tag{16}$$

Some further characterizations of normal L-subgroups are given below:

Theorem 1.13 Let $\eta \in L(\mu)$. Then, the following are equivalent:

i. $\eta \in NL(\mu)$,

ii. $a_x \circ \eta = \eta \circ a_x$ for each $a_x \in \mu$,

iii. $a_x \circ \eta \circ a_{x^{-1}} \subseteq \eta$ for each $a_x \in \mu$,

iv. $a_x \circ b_y \circ a_{x^{-1}} \in \eta$ for each $a_x \in \mu$ and $b_y \in \eta$.

The following fact is well known:

If $\mu \in L^X$, then $\mu = \bigcup_{x \in X} \{(\mu(x))_x\}$.

This immediately yields:

Theorem 1.14 Let $\eta, \theta \in L^G$. Then, $\eta \circ \theta = \bigcup_{x \in G} \{(\eta(x))_x \circ \theta\} = \bigcup_{x \in G} \{\eta \circ (\theta(x))_x\}$.

Hence we have that each L-subset of μ commutes with every normal L-subgroup of μ.

Theorem 1.15 Let $\eta \in NL(\mu)$. Then, $\eta \circ \theta = \theta \circ \eta$ for each $\theta \in L^{\mu}$.

The notion of normalizer which was so far introduced and discussed in fuzzy group theory is a crisp subset (subgroup) of the given parent group G. Moreover, this normalizer turns out to be the intersection of normalizers of all the level subsets (subgroups) of the fuzzy subgroup in question. Another drawback of this normalizer is that each member of a certain equivalence class of fuzzy subgroups of a fuzzy group has the same normalizer. This phenomenon arises due to the fact that the studies, carried out by these researchers are for the fuzzy subgroups of an ordinary group. Here we demonstrate, how we can introduce the concept of a normalizer which is not an ordinary subgroup but an L-subgroup itself and satisfies most of the properties of the notion of the normalizer of an ordinary subgroup of a group. Firstly, we present the construction of this notion in the following:

Theorem 1.16 Let $\eta \in L(\mu)$. Define an L-subset δ of G as follows:

$$\delta = \bigcup_{a_x \in \mu} \{a_x : a_x \circ \eta = \eta \circ a_x\}. \tag{17}$$

Then, δ is the largest L-subgroup of μ such that η is a normal L-subgroup of δ. Here δ is called the normalizer of η and is denoted by $N(\eta)$. Moreover, it turns out that $N(\eta)(e) = \mu(e)$.

This immediately leads us to the following result:

Let $\eta \in L(\mu)$. Then, $\eta \in NL(\mu)$ if and only if $N(\eta) = \mu$.

On the other hand, if we replace the parent L-group μ by 1_G, then we have:

Theorem 1.18 Let $\eta \in L(G)$. Then, f(η) is an normal L-subgroup of G if and only if $N(\eta) = 1_G$.

As a consequence, we recover the classical result:

Theorem 1.19 Let H be a subgroup of G. Then, $x \in N(H)$ if and only if $1_H \circ 1_x = 1_x \circ 1_H$.

Below we provide some more properties related to the normalizer so defined:

Theorem 1.20 Let $\eta, \theta \in L(\mu)$. Then

i. $N(\eta) \cap N(\theta) \subseteq N(\eta \cap \theta)$,

ii. $N(\eta) \cap N(\theta) \subseteq N(\eta \circ \theta)$ provided $\eta \circ \theta \in L(\mu)$.

Now, the following results reflect the behavior of this normalizer under the action of a group homomorphism. We start with:

Theorem 1.21 Let $f : G \to K$ be a group homomorphism and $x \in G$. If $y \in K$, then the set of all preimages of '$yf(x)$' is precisely the set of all elements of the form 'ux' where $f(u) = y$.

Theorem 1.22 Let $f(v)$ be a group homomorphism and $\eta, \theta \in L^{\mu}$. Then, $f(\eta \circ \theta) = f(\eta) \circ f(\theta)$.

Moreover,

Theorem 1.23 Let $f : G \to K$ be a group homomorphism and $\nu \in L(K)$.If $\theta \in L^{\nu}$, then $f^{-1}(\theta \circ b_{f(x)}) = f^{-1}(\theta) \circ b_x$.

The above results are helpful in establishing the following:

Theorem 1.24 Let $f : G \to K$ be a group homomorphism. Then, for $\mu \in L(G)$ and $\nu \in L(K)$

i. $f(N(\eta)) \subseteq N(f(\eta))$ for each $\eta \in L(\mu)$,

ii. $f^{-1}(N(\theta)) \subseteq N(f^{-1}(\theta)))$ for each $\theta \in L(\nu)$.

3.2 Generated *L*-subgroup of an *L*-group

Firstly, we recall the following results for generating an L-subgroup by a given L-subset from [25] and study its relationship with other notions of L-group theory.

Theorem 1.25 Let $\eta \in L^{\mu}$. Let $a_0 = \underset{x \in G}{\vee}\{\eta(x)\}$ and define an L-subset $\hat{\eta}$ of G by

$$\hat{\eta}(x) = \underset{a \leq a_0}{\vee}\{a : x \in \langle \eta_a \rangle\}. \tag{18}$$

Then, $\hat{\eta} \in L(\mu)$ and $\hat{\eta} = \langle \eta \rangle$. Moreover, $\langle \eta \rangle(e) = \vee_{x \in G}\{\eta(x)\}$.

The above theorem is used to establish the following:

Theorem 1.26 Let $\eta \in NL^{\mu}$. Then, $\langle \eta \rangle \in NL(\mu)$.

In the following, we demonstrate the significance of sup-property in the studies of L-group theory:

Theorem 1.27 Let $\eta \in L_s^{\mu}$. Then, define an L-subset $\hat{\eta}$ of G by

$$\hat{\eta}(x) = \underset{a \in Im\eta}{\vee}\{a : x \in \langle \eta_a \rangle\}. \tag{19}$$

Then, $\hat{\eta} \in L(\mu)$ and $\hat{\eta} = \langle \eta \rangle$. Moreover, $\hat{\eta}$ possesses sup-property and $\text{Im}\,\hat{\eta} \subseteq \text{Im}\,\eta$.

The following result is an immediate consequence of the above theorem:

Theorem 1.28 Let $\eta \in L_s^{\mu}$. If $a_0 = \underset{x \in G}{\vee}\{\eta(x)\}$, then for each $b \leq a_0$, $\langle \eta_b \rangle = \langle \eta \rangle_b$.

The following example will demonstrate that the condition of sup-property is crucial and can not be removed from the above result:

Example 1: Let Z be the group of integers under addition, and let $\langle 2^n \rangle$ be the subgroup of Z generated by 2^n, where n is a fixed positive integer. Then the direct product $Z \times Z$ contains subgroups

$$\langle 2^r \rangle \times \langle 2^s \rangle \text{ for each } r, s = 0,1,2, \ldots \tag{20}$$

Define the following L-subset of $Z \times Z$ where L is the closed unit interval ordered by usual ordering of real numbers:

$$\mu(x) = \begin{cases} 0 & \text{if } x \in Z \times Z \sim \langle 2 \rangle \times Z, \\ \frac{3}{4} & \text{if } x \in \langle 2 \rangle \times Z. \end{cases} \tag{21}$$

$$\eta(x) = \begin{cases} 0 & \text{if } x \in x \in Z \times Z \sim \langle 2 \rangle \times Z, \\ \frac{1}{2}\left(1 - \frac{1}{2^n}\right) & \text{if } x \in \langle 2^n \rangle \times Z \sim \langle 2^{n+1} \rangle \times Z, \text{ where } n = 1,2,3, \ldots \\ 0 & \text{if } x \in \langle 0 \rangle \times Z \sim \langle 0 \rangle \times \langle 2 \rangle, \\ \frac{3}{4}\left(1 - \frac{1}{4^n}\right) & \text{if } x \in \langle 0 \rangle \times \langle 2^n \rangle \sim \langle 0 \rangle \times \langle 2^{n+1} \rangle, \text{ where } n = 1,2,3, \ldots \\ \frac{3}{4} & \text{if } x = (0,0). \end{cases} \tag{22}$$

Here $A \sim B$ means usual set difference.Clearly, $\eta \subseteq \mu$, $\eta \neq \mu$ and $\mu \in L(G)$. Observe that η does not possess sup-property and for $t = \frac{1}{2}$,

$$\langle 0 \rangle \times \langle 2 \rangle = \left\langle \eta_{\frac{1}{2}} \right\rangle \subset \langle \eta \rangle_{\frac{1}{2}} = \langle 0 \rangle \times Z. \tag{23}$$

Moreover,

Theorem 1.29 Let $\eta \in L^{\mu}$ and $a_0 = \underset{x \in G}{\vee}\eta(x)$. If L is a chain, then $\left\langle \eta_a^{>} \right\rangle = \langle \eta \rangle_a^{>}$ for each $a < a_0$.

3.3 Characteristic L-subgroup of an L-group

The notion of a characteristic subgroup of a group has been extended to the fuzzy setting by many researchers in the past. However in all these attempts, the parent group in question is an ordinary group rather than a fuzzy group. Here in this section, we firstly introduce the notion of a characteristic L-subset of an L-group. Then, we introduce the notion of a characteristic L-subgroup of an L-group in a manner similar to that of a normal L-subgroup of an L-group introduced earlier. After providing its characterization in terms of level subsets, we provide some group theoretic analogs to establish this notion. We also prove that the set of characteristic L-subsets (subgroups) is closed under arbitrary intersections (see [27]).

Definition 1.9 Let $\eta \in L^{\mu}$ with tip a_0. Then, η is said to be a characteristic L-subset of μ if

$$\eta(Tx) \geq \eta(x) \text{ for each } T \in A(\mu_a) \text{ and for each } a \leq a_0; \tag{24}$$

where $A(\mu_a)$ is the group of automorphisms of μ_a. We denote the set of all characteristic L-subsets of μ by CL^{μ}.

It is easy to see that μ is a characteristic L-subset of itself.

Theorem 1.30 Let $\eta \in L^{\mu}$ with tip a_0. Then, $\eta \in CL^{\mu}$ if and only if η_a is a characteristic subset of μ_a for each $a \leq a_0$.

The set of all normal L-subsets of the L-group μ is denoted by NL^{μ}. The following results are immediate:

Theorem 1.31 Let $\eta \in CL^{\mu}$. Then, $\eta \in NL^{\mu}$.

Theorem 1.32 Let $\eta \in CL^{\mu}$ with tip a_0. Then,

$$T(\eta|\mu_a) = \eta|\mu_a \quad \text{for all } T \in A(\mu_a); \quad \text{where} \quad a \leq a_0. \tag{25}$$

The set of all L-subsets of μ possessing sup-property is denoted by L_s^{μ}. The following result has been discussed in the fuzzy setting in [27]:

Theorem 1.33 Let $\eta, \theta \in L_s^{\mu}$. Then, $\eta \cup \theta$ and $\eta \cap \theta \in L_s^{\mu}$ provided L is a chain.

Now, since the meet and the join operations in L^{μ} are defined to be the intersection and the union of L-subsets respectively, the set L_s^{μ} constitutes a sublattice of L^{μ} provided L is a chain.

It is easy to observe that if L is a chain, then any L-subgroup of an L-group with finite range possesses sup-property. The situation, however, in the case of infinite range is varied and interesting (see [46]). Also, we see the role of sup-property when L is not a chain. Here we present a generalization of the notion of sup-property in order to obtain certain results. The following characterization of sup-property forms the basis of our generalization:

Theorem 1.34 Let $\eta \in L^{\mu}$. Then, η possesses sup-property if and only if each non-empty subset of $Im\eta$ is closed under arbitrary supremum.

Definition 1.10 A non-empty subset X of a lattice L is said to be supstar if every non-empty subset A of X contains its supremum. That is, if $\sup A = a_0$, then $a_0 \in A$.

By the definition, it is clear that every subset of a supstar subset is again a supstar subset. Now, define:

Definition 1.11 Let $\{\eta_i\}_{i \in I} \subseteq L^{\mu}$ be an arbitrary family. Then, $\{\eta_i\}_{i \in I}$ is said to be a supstar family if $\bigcup_{i \in I} Im\eta_i$ is a supstar subset of L. In particular, a pair of L-subsets η and θ is said to be jointly supstar if the set $\{\eta, \theta\}$ is a supstar family.

The following results are obtained by using the notions of sup-property and supstar family:

Theorem 1.35 Let $\eta \in L^{\mu}$. Then, η has sup-property if and only if $Im\eta$ is a supstar subset of L.

Theorem 1.36 Let $\{\eta_i\}_{i \in I} \subseteq L^{\mu}$ be a supstar family. Then,

i. η_i possesses sup-property for each $i \in I$,

ii. $\bigcup_{i \in \Omega} \eta_i$ possess sup-property, where $\Omega \subseteq I$.

Moreover, we have:

Theorem 1.37 If $\{\eta_i\}_{i \in I} \subseteq L^{\mu}$ is a maximal supstar family, then $\{\eta_i\}_{i \in I}$ is a complete lattice under the ordering of L-set inclusion.

In order to study the lattice theoretic behavior of characteristic L-subsets, have the following:

Theorem 1.38 Let $\{\eta_i\}_{i \in I} \subseteq CL^{\mu}$. Then,

i. $\bigcap_{i \in I} \eta_i \in CL^{\mu}$,

ii. $\bigcup_{i \in I} \eta_i \in CL^{\mu}$ provided $\{\eta_i\}_{i \in I}$ is a supstar family.

In view of Theorem 1.31, $CL^{\mu} \subseteq NL^{\mu} \subseteq L^{\mu}$. By Theorem 1.6, NL^{μ} is closed under arbitrary unions and intersections. Hence NL^{μ} is a complete sublattice of L^{μ}. Further by Theorem 1.38, CL^{μ} is closed under arbitrary intersections with the greatest element μ. Thus CL^{μ} is a lower complete sublattice of L^{μ} and is a complete lattice in its own right. Moreover, if both CL^{μ} and NL^{μ} are supstar families and the lattice L is a chain, then by Theorem 1.31 and Theorem 1.36, $CL^{\mu} \subseteq NL^{\mu} \subseteq L_s^{\mu}$. By Theorem 1.38, CL^{μ} is closed under arbitrary unions with the least element identically zero function. So CL^{μ} is an upper complete sublattices of NL^{μ}. Similarly by Theorem 1.6 and Theorem 1.36, NL^{μ} is an upper complete sublattices of L_s^{μ}.

On the other hand, we discuss the behavior of set products of L-subsets when the given lattice is a chain or the two L-subsets are jointly supstar.

Theorem 1.39 Let $\eta, \nu \in L^{\mu}$. Then,

i. $(\eta \circ \nu)_a^{>} = \eta_a^{>} \nu_a^{>}$ for each $a \in L \sim \{1\}$, provided L is a chain,

ii. $(\eta \circ \nu)_a = \eta_a \nu_a$ for each $a \in L$, provided η and ν are jointly supstar.

This helps us in establishing the following:

Theorem 1.40 Let $\eta, \theta \in CL^{\mu}$ be jointly supstar. Then, $\eta \circ \theta \in CL^{\mu}$.

Below we study the notion of a characteristic L-subgroup of an L-group and its related properties:

Definition 1.12 Let $\eta \in L(\mu)$. Then, η is said to be a characteristic L-subgroup of μ if $\eta \in CL^{\mu}$. We denote by $CL(\mu)$ the set of all characteristic L-subgroups of μ.

Clearly, an L-group μ is a characteristic L-subgroup of itself. The following theorem characterizes the notion of a characteristic L-subgroup in terms of its level subsets:

Theorem 1.41 Let $\eta \in L(\mu)$. Then, $\eta \in CL(\mu)$ if and only if η_a is a characteristic subgroup of μ_a foreach $a \leq \eta(e)$.

Moreover, we have the following results:

Theorem 1.42 Let $\eta \in CL(\mu)$. Then, $\eta \in NL(\mu)$.

Theorem 1.43 Let $\eta \in CL(\mu)$. Then,

$$T(\eta|\mu_a)| = \eta|\mu_a \text{ for all } T \in A(\mu_a); \quad \text{where} \quad a \leq \eta(e). \tag{26}$$

Theorem 1.44 Let $\eta, \theta \in CL(\mu)$ be jointly supstar. Then, $\eta \circ \theta \in CL(\mu)$.

Theorem 1.45 Let $\theta \in NL(\mu)$ and $\eta \in CL(\theta)$. Then, $\eta \in NL(\mu)$.

In classical group theory, it is well known that the property of being a characteristic subgroup is transitive, and a characteristic subgroup of a normal subgroup of a group is a normal subgroup. However, the same could not even be formulated in the work of earlier researchers who have defined the notion of a characteristic fuzzy subgroup of an ordinary group in various ways. However, these pleasing features of classical group theory are retained in our studies.

Theorem 1.46 Let $\theta \in CL(\mu)$ and $\eta \in CL(\theta)$. Then, $\eta \in CL(\mu)$.

Let us denote by $CL_s(\mu)$ the set of all characteristic L-subgroups of μ, each member of which possesses sup-property. In classical group theory, the subgroup generated by a characteristic subset of a group is a characteristic subgroup. Below we provide its counterpart in L-group theory:

Theorem 1.47 Let $\eta \in CL^{\mu}$ and possesses sup-property. Then, $\langle \eta \rangle \in CL_s(\mu)$.

Now, we exhibit that the set of all normal L-subgroups and the set of all characteristic L-subgroups each member of which possesses sup-property, constitute sublattices of the lattice of L-subgroups of a given L-group.

Theorem 1.48 The set $NL(\mu)$ is a complete sublattice of $L(\mu)$..

In the following results the lattice L is a chain:

Theorem 1.49 The set $L_s(\mu)$ of all L-subgroups of μ each member of which possesses sup-property, is a sublattice of $L(\mu)$.

Theorem 1.50 The set $NL_s(\mu)$ of all normal L-subgroup of μ each member of which possesses sup-property, is a sublattice of $L_s(\mu)$ and hence of $L(\mu)$.

Theorem 1.51 The set $CL_s(\mu)$ of all characteristic L-subgroups of μ each member of which possesses sup-property, is a sublattice of $NL_s(\mu)$ and hence of $L(\mu)$.

The following diagram provides the lattice structure of the sublattices of the lattice $L(G)$ provided the lattice L is a chain (**Figure 1**):

If $L_t(\mu)$ denotes the set of all L-subgroups of μ each member of which has the same tip t, where $t \in Im\ \mu$, then the following result is easy to verify:

Theorem 1.52 The set $L_t(\mu)$ is a sublattice of $L(\mu)$.

Moreover, we have:

Theorem 1.53 The lattice $L(\mu)$ is a disjoint union of its sublattices $L_t(\mu)$. That is $L(\mu) = \bigcup_{t \in Im\ \mu} L_t(\mu)$.

As the intersection of two sublattices is a sublattice of the given lattice, the following results are immediate:

Theorem 1.54 The set $L_{nt}(\mu)$ of all normal L-subgroups of μ with the same tip t, is a sublattice of $L_n(\mu)$ and hence of $L(\mu)$.

In the following results the lattice L is a chain:

Theorem 1.55 The set $L_{st}(\mu)$ of all L-subgroups of μ with the same tip t and each member of which possesses sup-property, is a sublattice of $L_s(\mu)$ and hence of $L(\mu)$.

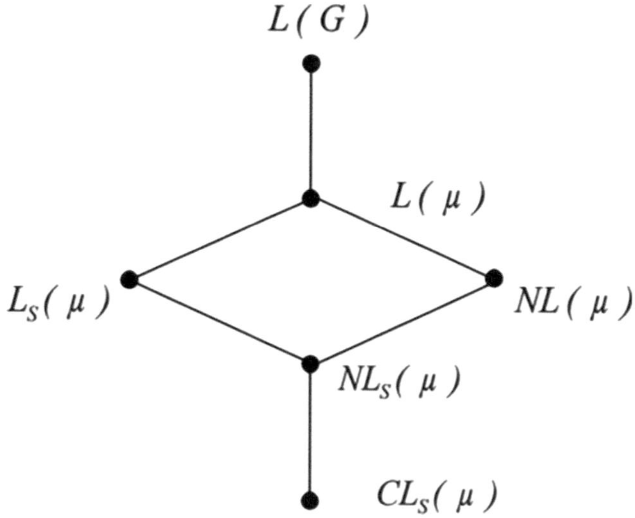

Figure 1.
The lattice structure of sublattices of the lattice $L(G)$.

Theorem 1.56 The set $L_{nst}(\mu)$ of all normal L-subgroups of μ with the same tip t and each member of which possesses sup-property, is a sublattice of $L_{nt}(\mu)$ and $L_{st}(\mu)$ and hence of $L(\mu)$.

The following lattice diagram shows the inter-relationship of the above discussed sublattices, where the lattice L is a chain (**Figure 2**):

Theorem 1.57 The set $L_{cst}(\mu)$ of all characteristic L-subgroups of μ with the same tip t and each member of which possesses sup-property, is a sublattice of $NL_{st}(\mu)$ and hence of $L(\mu)$ provided the lattice L is a chain.

Now, let us consider an arbitrary lattice L. Let $L^*(\mu)$ denote a subclass of the lattice $L(\mu)$ consisting of all the supstar families. Then, $L^*(\mu)$ is a complete sublattice of $L(\mu)$. Now if CL^*_{st}, $CL^*_s(\mu)$, $NL^*_s(\mu)$, $L^*_s(\mu)$, $NL^*_{st}(\mu)$ and $L^*_{st}(\mu)$ denote the set of all supstar families of L-subsets in CL_{st}, $CL_s(\mu)$, $NL_s(\mu)$, $L_s(\mu)$, $NL_{st}(\mu)$ and $L_{st}(\mu)$ respectively, then it follows that

$$CL^*_{st} \subseteq NL^*_{st} \subseteq L^*_{st} \subseteq L^*_s(\mu) \text{ and } CL^*_{st} \subseteq CL^*_s \subseteq NL^*_s \subseteq L^*_s(\mu); \tag{27}$$

are sublattices of $L(\mu)$. Note that $L^*_s(\mu) \subseteq L_s(\mu)$. Moreover, we mention that it is not known whether $CL(\mu)$ is a sublattice of $NL(\mu)$. We leave this question as an open problem and below we describe the inter-relationship of above discussed lattices, simply under the set inclusion, where the lattice L is taken to be a chain (**Figure 3**).

4. Commutator L-subgroup and nilpotent L-subgroup of an L-group

The class of nilpotent groups constitutes an important class in the studies of group theory. In fact, nilpotent groups are very near to Abelian groups and they are always solvable. They arise in the studies of Galois theory as well as in the classification of groups. In the investigation of nilpotent groups and solvable groups, the notion of commutator and commutator subgroups play very significant role. Therefore, we start with notion of commutator L-subgroups.

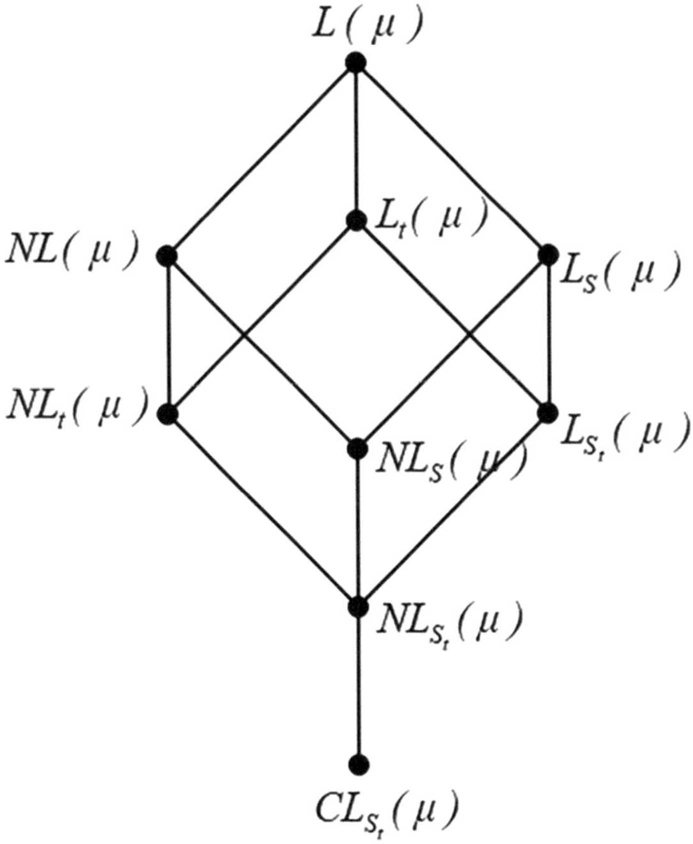

Figure 2.
Inter-relationship of the sublattices of $L(\mu)$.

4.1 Commutator L-subgroup of an L-group

In this subsection, we study the notion of a commutator in L-setting [28, 29] by using the notion of infimums of L-subsets. In fact, we also discuss here how infimums of L-subsets play an effective role in the development of the theory of L-subgroups.

Definition 1.13 Let $\eta, \theta \in L^\mu$. Then, the commutator of η and θ is an L-subset (η, θ) of G defined as follows:

$$(\eta,\theta)(x) = \begin{cases} \vee\{\eta(y)\wedge\theta(z)\}, & \text{if } x = [y,z] \text{ for some } y,z \in G, \\ \inf \eta \wedge \inf \theta, & \text{if } x \neq [y,z] \text{ for any } y,z \in G. \end{cases} \tag{28}$$

The commutator L-subgroup of $\eta, \theta \in L^\mu$ is defined as the L-subgroup of G generated by (η, θ). It is denoted by $[\eta, \theta]$. Thus, $[\eta, \theta] = \langle(\eta, \theta)\rangle$. Moreover, $[\eta, \theta](e) = \eta(e)\wedge\theta(e)$.

In general, if $\{\eta_i\}_{i=1}^n \subseteq L^\mu$, then we write $(\eta_1, \eta_2, \eta_3, \dots, \eta_n) = (((\eta_1, \eta_2), \eta_3), \dots, \eta_n)$ and $[\eta_1, \eta_2, \eta_3, \dots, \eta_n] = [[[\eta_1, \eta_2], \eta_3], \dots, \eta_n]$. Moreover, it follows that $\inf(\eta_1, \eta_2, \eta_3, \dots, \eta_n) = \inf \eta_1 \wedge \inf \eta_2 \wedge \dots \wedge \inf \eta_n$. Also, $[\eta_1, \eta_2, \eta_3, \dots \eta_n](e) = \eta_1(e)\wedge\eta_2(e)\wedge \dots \wedge\eta_n(e)$.

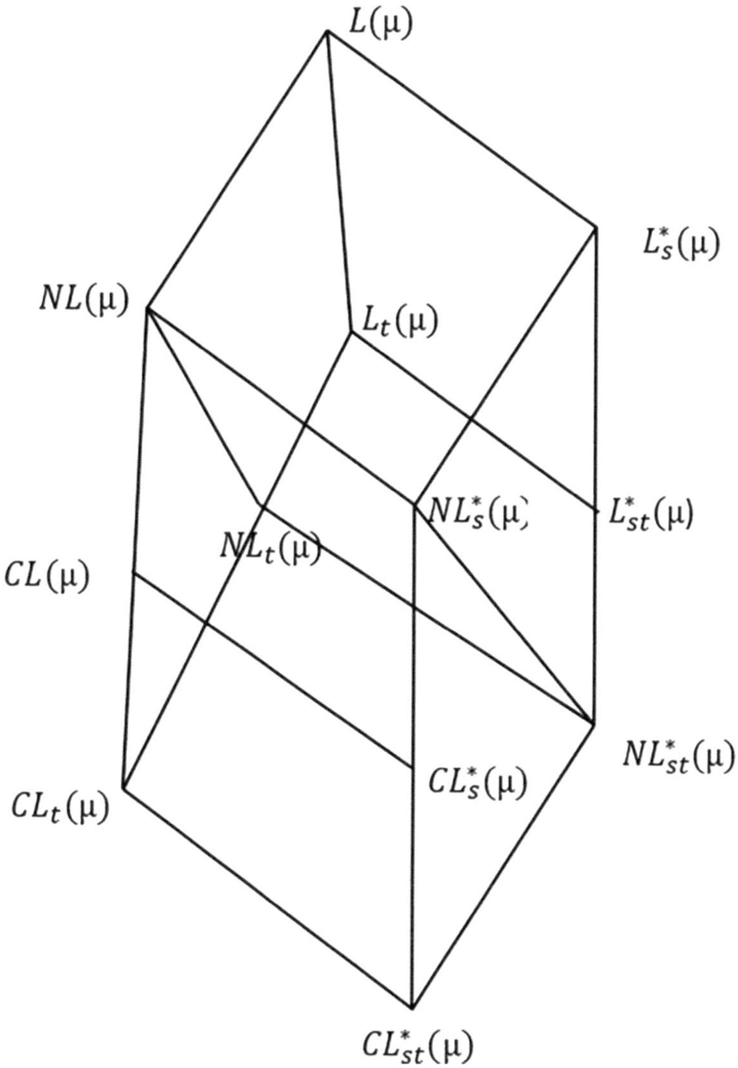

Figure 3.
Inter-relationship of various sublattices of $L(\mu)$ *under ordinary set inclusion.*

In the following, we present some extensions of the results of classical group theory with certain deviations and peculiarities:

Theorem 1.58 Let $\eta, \theta \in L^\mu$ and $\eta \subseteq \theta$. Then, $[\eta, \sigma] \subseteq [\theta, \sigma]$ for each $\sigma \in L^\mu$.

Theorem 1.59 Let $\eta, \theta \in L^\mu$. Then, $[\eta, \theta] = [\theta, \eta]$.

Theorem 1.60 Let $\eta, \theta \in L^\mu$. Then,

$$\inf \eta \wedge \inf \theta \leq \inf \eta \circ \theta \leq \inf \eta \vee \inf \theta. \tag{29}$$

Theorem 1.61 Let $\eta, \theta \in NL(\mu)$ and $\sigma \in L(\mu)$. If either η and θ or θ and σ have the same tails, then

$$[\sigma \circ \eta, \theta] \subseteq [\eta, \theta] \circ [\sigma, \theta], \tag{30}$$

and moreover, if $\eta(e) = \sigma(e)$, then the equality holds.

Theorem 1.62 Let $\eta, \theta \in NL(\mu)$. Then,

$$[\eta, \theta] \in NL(\mu) \text{ and } [\eta, \theta] \subseteq \eta \cap \theta. \tag{31}$$

Theorem 1.63 Let $\eta, \theta \in L^{\mu}$ and $f : G \to K$ be a group homomorphism. Then, $f(\langle \eta \rangle) = \langle f(\eta) \rangle$ and $f([\eta, \theta]) = \langle f((\eta, \theta)) \rangle$.

Theorem 1.64 Let $\eta, \theta \in CL^{\mu}$ be jointly supstar. Then, $[\eta, \theta] \in CL_s(\mu)$.

Theorem 1.65 Let $\eta, \theta \in L^{\mu}$ and $f : G \to K$ be a homomorphism. Let $\inf \eta = \inf f(\eta)$ and $\inf \theta = \inf f(\theta)$. Then, $[f(\eta), f(\theta)] = f([\eta, \theta])$.

Theorem 1.66 Let $f : G \to H$ be a homomorphism and $\nu \in L(H)$. Let $\lambda, \sigma \in L^{\nu}$ and $\inf \lambda = \inf f^{-1}(\lambda)$, $\inf \sigma = \inf f^{-1}(\sigma)$. Then, $[f^{-1}(\lambda), f^{-1}(\sigma)] \subseteq f^{-1}([\lambda, \sigma])$.

Theorem 1.67 Let $\eta, \theta \in L(\mu)$ be such that η and θ are jointly supstar. Then, the commutator (η, θ) and hence the commutator L-subgroup $[\eta, \theta]$ possess sup-property.

Note that, $a \leq \inf \eta$ if and only if $\eta_a = G$. Moreover, if $a \leq \inf \eta \wedge \inf \theta$, then the level subsets η_a, θ_a and $(\eta, \theta)_a$ coincide with G and hence $(\eta, \theta)_a \neq (\eta_a, \theta_a)$.

Theorem 1.68 Let $\eta, \theta \in L(\mu)$. If $a_0 = \eta(e) \wedge \theta(e)$ and $a \nleq \inf \eta \wedge \inf \theta$, then

i. $[\eta, \theta]_a = [\eta_a, \theta_a]$ for each $a \leq a_0$, provided η and θ are jointly supstar.

ii. $[\eta, \theta]_a^{>} = [\eta_a^{>}, \theta_a^{>}]$ for each $a < a_0$, provided L is a chain.

In general, let $\{\eta_i\}_{i=1}^{n} \subseteq L(\mu)$. If $a_0 = \eta_1(e) \wedge \eta_2(e) \wedge \ldots \wedge \eta_n(e)$ and $a \nleq \inf \eta_i$ for each i, then

i. $[\eta_1, \eta_2, \eta_3, \ldots, \eta_n]_a = [(\eta_1)_a, (\eta_2)_a, (\eta_3)_a, \ldots, (\eta_n)_a]$ for each $a \leq a_0$, provided $\{\eta_i\}_{i=1}^{n}$ is a supstar family.

ii. $[\eta_1, \eta_2, \eta_3, \ldots, \eta_n]_a^{>} = [(\eta_1)_a^{>}, (\eta_2)_a^{>}, (\eta_3)_a^{>}, \ldots, (\eta_n)_a^{>}]$ for each $a < a_0$, provided L is a chain.

4.2 Nilpotent L-subgroup of an L-group

We begin this subsection with the concept of a descending central chain of an L-subgroup by making use of the notion of commutators. Then, this in turn has been used to define nilpotent L-subgroups [28]. Throughout this subsection, G would denote a group which is not perfect.

We start with the definition of a descending central chain of an L-subgroup η of an L-subgroup μ.

Take $Z_0(\eta) = \eta$, $Z_1(\eta) = [Z_0(\eta), \eta]$. And in general, for each i, define

$$Z_i(\eta) = [Z_{i-1}(\eta), \eta]. \tag{32}$$

The following result is an immediate consequence of the above definition:

Theorem 1.69 Let $\eta \in L(\mu)$. Then for each $i, Z_i(\eta) \subseteq Z_{i-1}(\eta)$.

Here we provide the definition of a descending central chain.

Definition 1.14 Let $\eta \in L(\mu)$. Then, the chain

$$\eta = Z_0(\eta) \supseteq Z_1(\eta) \supseteq \cdots \supseteq Z_i(\eta) \supseteq \cdots \tag{33}$$

of L-subgroups of μ is called the descending central chain of η.

It is worthwhile to note that as $\eta \in NL(\eta)$, in view of Theorem 1.62, $Z_i(\eta)$ is a normal L-subgroup of η for each i. Moreover, if $\eta \in L(\mu)$, then $Z_i(\eta) \in NL(\mu)$.

Now, we are in a position to formulate the definition of a nilpotent L-subgroup of an L-group.

Definition 1.15 Let $\eta \in L(\mu)$ with tip a_0 and tail t_0 and $a_0 \neq t_0$. If the descending central chain

$$\eta = Z_0(\eta) \supseteq Z_1(\eta) \supseteq \cdots \supseteq Z_i(\eta) \supseteq \cdots \tag{34}$$

terminates finitely to the trivial L-subgroup $\eta_{t_0}^{a_0}$, then η is known as a nilpotent L-subgroup of μ. More precisely, η is said to be nilpotent of class c if c is the least non-negative integer such that $Z_c(\eta) = \eta_{t_0}^{a_0}$. In this case, the series

$$\eta = Z_0(\eta) \supseteq Z_1(\eta) \supseteq \cdots \supseteq Z_c(\eta) = \eta_{t_0}^{a_0} \tag{35}$$

is called the descending central series of η. If it is a nilpotent L-subgroup of μ, then we simply write η is nilpotent. Clearly, the tip and tail of the members $Z_i(\eta)$ of descending central series coincide with the tip and the tail of the trivial L-subgroup $\eta_{t_0}^{a_0}$.

Next, we provide some results pertaining to the member '$Z_i(\eta)$' of the descending central chain which are peculiarities of L-setting.

Theorem 1.70 Let $\eta \in L(\mu)$ and possesses sup-property. Then for each i,

$$Im Z_i(\eta) \subseteq Im\eta \cup \{\inf \eta\}. \tag{36}$$

Theorem 1.71 Let $\eta \in L(\mu)$ and possesses sup-property. Then for each i,

i. $Z_i(\eta)$ possesses sup-property.

ii. $Z_i(\eta)$ and η are jointly supstar.

The following result justifies the naturality of the extension of the notion of descending central chain:

Theorem 1.72 Let H be a subgroup of G. Then for each i, $Z_i(1_H) = 1_{Z_i(H)}$.

For further justification of these notions, we provide the level subset and strong level subset characterizations of the members of descending central series.

Theorem 1.73 Let $\eta \in L(\mu)$ and possesses sup-property. Then for each $a \nleq \inf \eta$ and $a \leq \eta(e)$, $Z_i(\eta_a) = (Z_i(\eta))_a$ for each i.

Theorem 1.74 Let L be a chain and $\eta \in L(\mu)$. Then for each a, where $\inf \eta \leq a < \eta(e)$, $Z_i(\eta_a^>) = (Z_i(\eta))_a^>$ for each i.

Finally, we obtain the level subset and strong level subset characterizations for nilpotent L-subgroups.

Theorem 1.75 Let $\eta \in L(\mu)$ and possesses sup-property. Then, η is a nilpotent L-subgroup of μ of nilpotent length at most n if and only if η_a is a nilpotent subgroup of μ_a of nilpotent length at most n for each $a \not\leq \inf \eta$ and $a \leq \eta(e)$.

Theorem 1.76 Let $\eta \in L(\mu)$ and L be a chain. Then, η is a nilpotent L-subgroup of μ of nilpotent length at most n if and only if $\eta_a^>$ is a nilpotent subgroup of $\mu_a^>$ of nilpotent length at most n, for each a, where $\inf \eta \leq a < \eta(e)$.

The following result immediately follows from the above results:

Theorem 1.77 A subgroup H of a group of G is nilpotent if and only if 1_H is a nilpotent L-subgroup of 1_G.

Next, we provide the definition of a central chain and use it to characterize nilpotent L-subgroups.

Definition 1.16 Let $\eta \in L(\mu)$ with tip a_0 and tail t_0. Then, the chain

$$\eta = \eta_0 \supseteq \eta_1 \supseteq \dots \supseteq \eta_n \supseteq \dots \tag{37}$$

of L-subgroups of μ is called a central chain of η, if for each i, $[\eta_{i-1}, \eta] \subseteq \eta_i$. If $a_0 \neq t_0$ and there exists a positive integer m such that $\eta_m = \eta_{t_0}^{a_0}$, where $\eta_{t_0}^{a_0}$ is the trivial L-subgroup of η with tip a_0 and tail t_0, then

$$\eta = \eta_0 \supseteq \eta_1 \supseteq \dots \supseteq \eta_m = \eta_{t_0}^{a_0} \tag{38}$$

is known as a central series of η. It follows that η_i and η have identical tips and also, identical tails for each i. Moreover, the following is easy to verify:

i. $[\eta_{i-1}, \eta] \subseteq \eta_i$ if and only if $\eta_i([x,y]) \geq \eta_{i-1}(x) \wedge \eta(y)$ for each i,

ii. $\eta_i \in NL(\eta)$ for each i.

Now, we are in a position to study the notion of a nilpotent L-subgroup by making use of the concept of a central chain. The results are as follows:

Theorem 1.78 Let $\eta \in L(\mu)$ be a proper L-subgroup of μ. Then, η is nilpotent if and only if η has a central series.

Theorem 1.79 Let $\eta \in L(\mu)$ and θ be a proper L-subgroup of η such that η and θ have the common tail t_0. If η is nilpotent, then θ is also nilpotent.

The notion of set product is an extension of the notion of product of complexes in classical group theory. The following two results provides a necessary mechanism for the set product of two nilpotent L-subgroups of μ to be nilpotent:

Theorem 1.80 Let $\eta, \eta_1, \dots, \eta_{n+1} \in NL(\mu)$ having identical tails.
If $\eta_i = \eta$ for k + 1 distinct values of i where $0 \leq k \leq n$, then $[\eta_1, \eta_2, \dots, \eta_{n+1}] \subseteq Z_k(\eta)$.

Theorem 1.81 Let η and θ be trivial L-subgroups of μ. Then, the set product $\eta \circ \theta$ is also a trivial L-subgroup of μ defined by

$$\eta \circ \theta(x) = \begin{cases} \eta(e) \wedge \theta(e) & \text{if } x = e, \\ inf\,\eta \vee inf\,\theta & \text{if } x \neq e, \end{cases} \tag{39}$$

if and only if $inf\,\eta \vee inf\,\theta < \eta(e) \wedge \theta(e)$.

The final result of this subsection provides a sufficient condition for the set product of two nilpotent L-subgroups to be nilpotent.

Theorem 1.82 Let $\eta, \theta \in \mathrm{NL}(\mu)$ with common tails t_0, such that $t_0 < \eta(e) \wedge \theta(e)$ and $\inf \eta \circ \theta = t_0$. If η and θ are nilpotent of classes c and d respectively, then $\eta \circ \theta$ is a nilpotent L-subgroup of μ of class at most $c + d$.

5. Solvable and supersolvable L-subgroups of an L-group

The significance of the notion of solvability in classical group theory is beyond doubt. The studies pertaining to the notion of solvability and supersolvability frequently occur in the literature. The application of solvability in Galois Theory is established. Also, the application of nilpotency and solvability are well known in Lie groups.

5.1 Solvable L-subgroups of an L-group

In this subsection, we study the notion of derived series of an L-subgroup of an L-group in the same fashion as in classical group theory (see [29]). Here, also, G would denote a group that is not perfect.

Let $\eta \in L(\mu)$. We define inductively the following sequence of L-subgroups of μ:

$$\eta^{(0)} = \eta \text{ and } \eta^{(i)} = \left[\eta^{(i-1)}, \eta^{(i-1)}\right] \text{ for each } i. \tag{40}$$

The following result is immediate:

Theorem 1.83 Let $\eta \in L(\mu)$. Then, $\eta^{(i)} \subseteq \eta^{(i-1)}$.

We study the concept of solvable L-subgroups of an L-group like its classical counterpart. For this purpose, we introduce the concept of a derived chain of an L-subgroup as follows:

Definition 1.17 Let $\eta \in L(\mu)$. Then, the chain

$$\eta = \eta^{(0)} \supseteq \eta^{(1)} \supseteq \cdots \supseteq \eta^{(i)} \supseteq \cdots \tag{41}$$

of L-subgroups of η is called the derived chain of η. Clearly, the tip of $\eta^{(i)}$ coincides with $\eta(e)$. Also, $\eta^{(i)} \in \mathrm{NL}(\eta)$. Moreover, if $\eta \in \mathrm{NL}(\mu)$, then $\eta^{(i)} \in \mathrm{NL}(\mu)$.

Now, we are in a position to formulate the definition of a solvable L-subgroup of an L-group.

Definition 1.18 Let $\eta \in L(\mu)$ with tip a_0 and tail t_0 and $a_0 \neq t_0$. If the derived chain

$$\eta = \eta^{(0)} \supseteq \eta^{(1)} \supseteq \cdots \supseteq \eta^{(i)} \supseteq \cdots \tag{42}$$

terminates finitely to the trivial L-subgroup $\eta_{t_0}^{a_0}$, then η is known as a solvable L-subgroup of μ. If m is the least non negative integer such that $\eta^{(m)} = \eta_{t_0}^{a_0}$, then the series

$$\eta = \eta^{(0)} \supseteq \eta^{(1)} \supseteq \cdots \supseteq \eta^{(m)} = \eta_{t_0}^{a_0} \tag{43}$$

is called the derived series of η and m is said to be the solvable length of η. If η is a solvable L-subgroup of μ, then we simply write η is solvable. Clearly, the tip and tail of the members $\eta^{(i)}$ of derived series coincide with the tip and the tail of the trivial L-subgroup $\eta_{t_0}^{a_0}$.

Next, we provide some results pertaining to the member'$\eta^{(i)}$' of the derived chain which are peculiarities of L-setting.

Theorem 1.84 Let $\eta \in L(\mu)$ and possesses sup-property. Then,

$$Im\, \eta^{(i)} \subseteq Im\, \eta \cup \{inf\ \eta\}. \tag{44}$$

Theorem 1.85 Let $\eta \in L(\mu)$ and possesses sup-property. Then,

i. $\eta^{(i)}$ possesses sup-property,

ii. $\eta^{(i)}$ and η are jointly supstar.

The following result justifies the naturality of the extension of the notion of derived chain:

Theorem 1.86 Let H be a subgroup of G. Then for each i, $(1_H)^{(i)} = 1_{H^{(i)}}$.

For the justification of these notions, we also provide the level subset and strong level subset characterizations of the members of derived series.

Theorem 1.87 Let $\eta \in L(\mu)$ and possesses sup-property. Then, for each $a \nleq inf\,\eta$ and $a \leq \eta(e)$, $\left(\eta^{(i)}\right)_a = (\eta_a)^{(i)}$ for each i.

Theorem 1.88 Let L be a chain and $\eta \in L(\mu)$. Then for each a, where $\inf \eta \leq a < \eta(e)$, $\left(\eta^i\right)_a^> = \left(\eta_a^>\right)^i$.

Finally, we obtain the level subset and strong level subset characterizations for solvable L-subgroups.

Theorem 1.89 Let $\eta \in L(\mu)$ and possesses sup-property. Then, η is a solvable L-subgroup of μ of solvable length at most n if and only if η_a is a solvable subgroup of μ_a of solvable length at most n for each $a \nleq inf\;\eta$ and $a \leq \eta(e)$.

Theorem 1.90 Let $\eta \in L(\mu)$ and L be a chain. Then, η is a solvable L-subgroup of μ of solvable length at most n if and only if $\eta_a^>$ is a solvable subgroup of $\mu_a^>$ of solvable length at most n for each a, where $\inf \eta \leq a < \eta(e)$.

In view of the above results, the following is immediate:

Theorem 1.91 A subgroup H of a group of G is solvable if and only if 1_H is a solvable L-subgroup of 1_G.

Now, we provide the definition of a solvable series and use it to characterize solvable L-subgroups.

Definition 1.19 Let $\eta \in L(\mu)$ be a proper L-subgroup with tip a_0 and tail t_0. If $\eta_{t_0}^{a_0}$ is the trivial L-subgroup of η, then a series

$$\eta = \eta_0 \supseteq \eta_1 \supseteq \cdots \supseteq \eta_n = \eta_{t_0}^{a_0} \tag{45}$$

of L-subgroups of η is said to be a solvable series for η, if for each i

$$[\eta_{i-1}, \eta_{i-1}] \subseteq \eta_i. \tag{46}$$

It follows that for each i, η_i and η have identical tips as well as identical tails. Moreover, the following is easy to verify:

(i) $[\eta_{i-1}, \eta_{i-1}] \subseteq \eta_i$ if and only if $\eta_i([x,y]) \geq \eta_{i-1}(x) \wedge \eta_{i-1}(y)$ for each i,

(ii) $\eta_i \in NL(\eta_{i-1})$ for each i.

The following is characterization relates the concept of solvability with that of solvable series:

Theorem 1.92 Let $\eta \in L(\mu)$ be a proper L-subgroup with tip a_0 and tail t_0. Then, η is solvable if and only if η has a solvable series.

The above characterization helps us in obtaining the following:

Theorem 1.93 Let $\eta, \theta \in L(\mu)$ be proper L-subgroups having identical tails. If η is a solvable L-subgroup and $\theta \subseteq \eta$, then θ is also solvable.

Our next result establishes the fact that every nilpotent L-subgroup is solvable.

Theorem 1.94 Let $\eta \in L(\mu)$ be a proper L-subgroup. Then, every central series of η is a solvable series.

In classical group theory, every nilpotent group is solvable. We obtain the same result in L-setting:

Theorem 1.95 Let $\eta \in L(\mu)$ be a proper L-subgroup. If η is nilpotent, then η is also solvable.

The nature of image and pre-image of a solvable L-subgroup under a group homomorphism is ascertained in the following results:

Theorem 1.96 Let $f : G \to H$ be a group homomorphism and $\eta \in L(\mu)$. Let $inf\ \eta = inf\ f(\eta)$. If η is solvable, then $f(\eta)$ is also solvable.

Theorem 1.97 Let $f : G \to H$ be a group homomorphism having solvable kernel and $\nu \in L(H)$. Let $\rho \in L(\nu)$ and $inf\ \rho = inf\ f^{-1}(\rho)$. If ρ is solvable, then $f^{-1}(\rho)$ is also solvable.

So far in our studies, we have not dealt with normal series and subinvariant series in L-setting. Here we introduce these concepts and utilize them to characterize solvable L-subgroups (see [32]). We start with:

Definition 1.20 Let $\eta \in L(\mu)$ be a proper L-subgroup with tip a_0 and tail t_0. Then, a sequence $\lambda_0, \lambda_1, \ldots, \lambda_n$ of L-subgroups of μ is said to be a normal (subinvariant) series of η if

$$\eta = \lambda_0 \supseteq \lambda_1 \supseteq \ldots \supseteq \lambda_n = \eta_{t_0}^{a_0} \tag{47}$$

and $\lambda_i \in NL(\mu)(\lambda_i \in NL(\lambda_{i-1}))$ for each i.

Remark: If $\lambda_0, \lambda_1, \ldots, \lambda_n$ is a normal (subinvariant) series of η, then $\lambda_i(e) = \eta(e) = a_0$ and $\inf \lambda_i = \inf \eta = t_0$ for each i.

Definition 1.21 Let $\eta \in L(\mu)$ be a proper L-subgroup with tip a_0 and tail t_0. Then, a normal (subinvariant) series $\lambda_0 = \eta \supseteq \lambda_1 \supseteq \ldots \supseteq \lambda_n = \eta_{t_0}^{a_0}$ of η is said to be a normal (subinvariant) series with Abelian factors if for each i, the factor group $\frac{(\lambda_{i-1})_a}{(\lambda_i)_a}$ is Abelian for each $t_0 < a \leq a_0$.

The following theorem extends a famous result of classical group theory pertaining to the notion of solvability to L-setting:

Theorem 1.98 Let L be a dense chain and $\eta \in L(\mu)$ be a proper L-subgroup with tip a_0 and tail t_0. Then, the following are equivalent:

i. η is solvable,

ii. η has a normal series with Abelian factors,

iii. η has a subinvariant series with Abelian factors.

Below we propose a definition of supersolvable subgroups in L-setting:

Definition 1.22 Let $\eta \in L(\mu)$ be a proper L-subgroup with tip a_0 and tail t_0. Then, η is said to be a supersolvable L-subgroup if η has a normal series $\lambda_0 = \eta \supseteq \lambda_1 \supseteq \ldots \supseteq \lambda_n = \eta_{t_0}^{a_0}$ such that each factor group $\frac{(\lambda_{i-1})_a}{(\lambda_i)_a}$ for each a, where $t_0 < a \leq a_0$, is cyclic.

It has been established earlier, that L-subgroups of nilpotent L-subgroups and solvable L-subgroups are nilpotent and solvable respectively. It is well known that the class of supersolvable groups is also closed under the formation of subgroups in classical group theory. Here, we extend this property of supersovability to L-group theory.

Theorem 1.99 Let $\eta \in L(\mu)$ be a proper L-subgroup with tip a_0 and tail t_0. Let $\theta \in L(\eta)$ be such that η and θ have identical tips and also identical tails. If η is supersolvable, then θ is also supersolvable.

The nilpotency of the commutator of a supersolvable subgroup is also retained in L-setting. The result is as follows:

Theorem 1.100 Let L be an upper well ordered chain and $\eta \in L(\mu)$ be a proper L-subgroup with tip a_0 and tail t_0. If η is supersolvable, then the commutator L-subgroup $[\eta, \eta]$ is nilpotent.

5.2 Zassenhaus theorem and Schreier refinement theorem

In the earlier subsection, we have already introduced the concept of a normal (subinvariant) series of an L-subgroup of an L-group. In classical group theory, two normal series of a group are said to be equivalent if it is possible to set up a one to one correspondence between the factors of two series such that the paired factors are isomorphic. This is obtained by a certain type of factorization of a group into factor groups. Then, a generalization of second isomorphism theorem which is called Zassenhaus Theorem is used to establish this fact. Here in L-setting for the equivalence of two normal (subinvariant) series, we consider the factorization of each level of the L-subgroups in the spirit of classical group theory. Then, we extend Zassenhaus Theorem to L-setting and utilize it to establish a version of Schreier Refinement Theorem.

Below we extend the definitions of a refinement of a normal (subinvariant) series and equivalent normal (subinvariant) series (see [32]):

Definition 1.23 Let $\eta \in L(\mu)$ with tip a_0 and tail t_0 such that $a_0 \neq t_0$. A normal (subinvariant) series of η

$$\eta = \theta_0 \supseteq \theta_1 \supseteq \theta_2 \supseteq \dots \supseteq \theta_m = \eta_{t_0}^{a_0} \tag{48}$$

is said to be a refinement of a normal (subinvariant) series of η

$$\eta = \eta_0 \supseteq \eta_1 \supseteq \eta_2 \supseteq \dots \supseteq \eta_n = \eta_{t_0}^{a_0} \tag{49}$$

if $\eta_0, \eta_1, \eta_2, \dots, \eta_n$ is a subsequence of $\theta_0, \theta_1, \theta_2, \dots, \theta_m$.

Definition 1.24 Let $\eta \in L(\mu)$ with tip a_0 and tail t_0 such that $a_0 \neq t_0$. Then, two normal (subinvariant) series $\eta = \eta_0 \supseteq \eta_1 \supseteq \eta_2 \supseteq \dots \supseteq \eta_n = \eta_{t_0}^{a_0}$ and $\eta = \theta_0 \supseteq \theta_1 \supseteq \theta_2 \supseteq \dots \supseteq \theta_m = \eta_{t_0}^{a_0}$ of an L-subgroup η are said to be equivalent if for each fixed $a \leq a_0$, the factors of the series

$$\eta_a = (\eta_0)_a \supseteq (\eta_1)_a \supseteq (\eta_2)_a \supseteq \dots \supseteq (\eta_m)_e = \{e\} \tag{50}$$

can be put in one to one correspondence with the factors of the series

$$\eta_a = (\theta_0)_a \supseteq (\theta_1)_a \supseteq (\theta_2)_a \supseteq \dots \supseteq (\theta_m)_e = \{e\} \tag{51}$$

in such a way that the paired factors are isomorphic.

Now, we extend Zassenhaus Theorem to L-setting:

Theorem 1.101 (Zassenhaus Theorem) Let $\eta, \theta \in L(\mu)$. Let $\eta_1 \in NL(\eta)$ and $\theta_1 \in NL(\theta)$. Then,

$$(\eta\cap\theta_1)\circ\eta_1\lhd(\eta\cap\theta)\circ\eta_1 \text{ and } (\eta_1\cap\theta)\circ\theta_1\lhd(\eta\cap\theta)\circ\theta_1. \tag{52}$$

Also, there is an isomorphism such that

$$\frac{(\eta\cap\theta)_a(\eta_1)_a}{(\eta\cap\theta_1)_a(\eta_1)_a} \cong \frac{(\eta\cap\theta)_a(\theta_1)_a}{(\eta_1\cap\theta)_a(\theta_1)_a} \text{ for each } a\leq a_0; \tag{53}$$

where $a_0 = \eta_1(e)\wedge\theta_1(e)$.

The following theorem provides an extension of Schreier Refinement Theorem to L-setting:

Theorem 1.102 (Schreier Refinement Theorem) Let L be an upper well ordered chain. Then, every two normal series of an L-subgroup have refinements that are equivalent.

6. Normal closure and subnormal L-subgroups of an L-group

The study of a normal closure is important in classical group theory. The concept arises due to the fact that certain subgroups of a group are away from being normal. It is the the smallest subgroup containing a given subgroup which is normal in the group. This notion leads to some refined concepts in classical group theory such as normal closure series and subnormality.

6.1 Normal closure series and subnormality

In order to define a suitable notion of normal closure in L-setting which can be used to introduce the notion of subnormality, we start with the following (see [30]):

Definition 1.25 Let $\eta\in L(\mu)$. Define an L-subset $\mu\eta\mu^{-1}$ of G as follows:

$$\mu\eta\mu^{-1}(x) = \bigvee_{\substack{x=zyz^{-1}\\ y,z\in G}} \{\eta(y)\wedge\mu(z)\} \quad \text{for each } x\in G. \tag{54}$$

We call the L-subset $\mu\eta\mu^{-1}$, conjugate of η in μ. Clearly, $\eta\subseteq\mu\eta\mu^{-1}\subseteq\mu$. Moreover, $\bigvee_{x\in G}\{\mu\eta\mu^{-1}(x)\} = \eta(e)$ and $\mu\eta\mu^{-1}(x) = \mu\eta\mu^{-1}(x^{-1})$ for each $x\in G$. The normal closure of η in μ is defined as the L-subgroup of μ generated by the conjugate $\mu\eta\mu^{-1}$. It is denoted by η^μ. Thus $\eta^\mu = \langle\mu\eta\mu^{-1}\rangle$.

The above defined notion satisfies the characteristic properties of a normal closure and retains the usual group theoretic relationship with the concepts of commutator subgroups and set product in L-setting. The result are as follows:

Theorem 1.103 Let $\eta\in L(\mu)$. Then, η^μ is the least normal L-subgroup of μ containing η.

Theorem 1.104 Let $\eta\in L(\mu)$. Then, $[\mu,\eta]\circ\eta\in L(\mu)$.

Theorem 1.105 Let $\eta,\theta\in L(\mu)$. Then,

i. $[\eta,\theta]\subseteq\eta^\mu\cap\theta^\mu$,

ii. $\eta^\mu = \eta\circ[\eta,\mu]$.

Before we embark on the study of normal closure series and subnormality in L-setting, we generalize notion of conjugacy and provide the definition of a conjugate of an L-subset by an L-subset (see [10, 33]).

Definition 1.26 Let $\eta, \theta \in L^{\mu}$. Define an L-subset $\theta\eta\theta^{-1}$ of G as follows:

$$\theta\eta\theta^{-1}(x) = \bigvee_{x=zyz^{-1}} \{\eta(y) \wedge \theta(z)\} \quad \text{for each } x \in G. \tag{55}$$

We call the L-subset $\theta\eta\theta^{-1}$ the conjugate of η by θ. Clearly, $\theta\eta\theta^{-1} \subseteq \mu$. Hence the L-subgroup $\langle \theta\eta\theta^{-1} \rangle \in L(\mu)$ and is denoted by η^{θ}.

Following theorem is instrumental in the development of this subsection:

Theorem 1.106 Let $\eta, \theta \in L(\mu)$. Then,

i. $\eta \subseteq \theta\eta\theta^{-1}$ provided $\eta(e) \leqslant \theta(e)$,

ii. $\bigvee_{x \in G} \{\theta\eta\theta^{-1}(x)\} = \theta\eta\theta^{-1}(e) = \eta(e) \wedge \theta(e)$,

iii. $\theta\eta\theta^{-1}(x) = \theta\eta\theta^{-1}(x^{-1})$ for each $x \in G$,

iv. $\theta\eta\theta^{-1}(gxg^{-1}) \geq \theta\eta\theta^{-1}(x) \wedge \theta(g)$ for each $x, g \in G$.

Firstly, we discuss here some properties of conjugate of L-subsets where the L-subsets in question are L-subgroups. The significance of such properties have already been shown in classical group theory for establishing certain properties of ith normal closure of a subgroup of a group.

Theorem 1.107 Let $\theta \subseteq \gamma$ and $\eta(e) = \theta(e)$. Then,

Let $\eta, \theta, \gamma \in L(\mu)$.

$$(\eta^{\gamma})^{\theta} = \eta^{\gamma}, \tag{56}$$

$$(\eta^{\theta})^{\gamma} = \eta^{\gamma}, \tag{57}$$

$$\eta^{\theta \circ \eta} = \eta^{\theta}. \tag{58}$$

Theorem 1.08 Let $\eta, \theta, \gamma \in L(\mu)$ be such that $\gamma(e) = \eta(e) = \theta(e)$. Then, $(\eta^{\theta})^{\gamma} = \eta^{\gamma \circ \theta}$.

Theorem 1.09 Let $\eta, \theta \in L(\mu)$ and $\eta^{\theta} = \eta$. Then, $\eta \circ \theta \in L(\mu)$.

In order to introduce the notion of subnormality of an L-subgroup of an L-group, we define a descending series. For $\eta \in L(\mu)$, define a series of L-subgroups of μ inductively as follows:

$$\eta_0 = \mu, \;\; \eta_1 = \eta^{\mu}, \;\; \eta_2 = \eta^{\eta_1}, \; \ldots, \eta_i = \eta^{\eta_{i-1}} \ldots \tag{59}$$

By Theorem 1.103, η_1 is the smallest normal L-subgroup of μ containing η and η_2 is the smallest normal L-subgroup of η_1 containing η and so on.Thus, we have

$$\eta \subseteq \cdots \lhd \eta_{i+1} \lhd \eta_i \lhd \cdots \lhd \eta_1 \lhd \eta_0 = \mu. \tag{60}$$

This inductively defined series is known as the normal closure series of η in μ and we call η_i the ith normal closure of η in μ. It is easy to verify that $\eta_i(e) = \eta(e)$ for each i.

Theorem 1.110 Let $\eta, \theta \in L(\mu)$ such that $\eta(e) = \theta(e)$. Let η_i be the ith normal closure of η in μ and $\eta_i^{\theta} = \eta_i$. Then, $\eta_{i+1} \in NL(\eta_i \circ \theta)$.

Now, define the notion of a subnormal L-subgroup of an L-group as follows:

Definition 1.27 Let $\eta \in L(\mu)$ and η_i be the *ith* normal closure of η in μ. If there exists a non negative integer m such that

$$\eta_m = \eta \triangleleft \eta_{m-1} \triangleleft \cdots \triangleleft \eta_0 = \mu, \tag{61}$$

then η is known as a subnormal L-subgroup of μ with defect m. We shall denote a subnormal L-subgroup η of μ with defect m by $\eta \overset{m}{\triangleleft} \mu$. If η is a subnormal L-subgroup of μ, then we shall write η is subnormal in μ.

Remark: Obviously m equals 0 if $\eta = \mu$ and $m = 1$ if $\eta \in NL(\mu)$ and $\eta \neq \mu$.

The following theorem shows that the normal closure series is the fastest descending normal series [33]:

Theorem 1.111 Let $\eta \in L(\mu)$ and

$$\eta \subseteq \cdots \triangleleft \eta_{i+1} \triangleleft \eta_i \triangleleft \cdots \triangleleft \eta_1 \triangleleft \eta_0 = \mu \tag{62}$$

be the normal closure series of η. If there exists a descending series $\gamma_0 = \mu, \gamma_1, \ldots, \gamma_i \ldots$ of L-subgroups of μ such that

$$\eta \subseteq \ldots \triangleleft \gamma_{i+1} \triangleleft \ldots \gamma_1 \triangleleft \gamma_0 = \mu, \tag{63}$$

then $\eta_i \subseteq \gamma_i$.

Following is the definition of a subnormal series of an L-group:

Definition 1.28 Let $\eta \in L(\mu)$. A finite series $\theta_0 = \mu, \theta_1, \theta_2, \ldots, \theta_m = \eta$ of L-subgroups of μ such that

$$\eta = \theta_m \triangleleft \theta_{m-1} \triangleleft \ldots \triangleleft \theta_0 = \mu \tag{64}$$

is said to be a subnormal series of η..

We shall describe the notion of a subnormal L-subgroup through the notion of above defined subnormal series. The following result inter-connects these two concepts:

Theorem 1.112 Let $\eta \in L(\mu)$. Then, η is a subnormal L-subgroup of μ having defect m if and only if η has a subnormal series

$$\eta = \gamma_m \triangleleft \ldots \triangleleft \gamma_{i+1} \triangleleft \ldots \gamma_1 \triangleleft \gamma_0 = \mu, \tag{65}$$

of length m and m is the smallest length of such a subnormal series.

The results given below are established with the help of the above theorem:

Theorem 1.113 Let η be a subnormal L-subgroup of μ with defect m.

i. Let $\theta \in L(\mu)$. Then, $\eta \cap \theta$ is a subnormal L-subgroup of θ with defect c where $c \leq m$. In particular, η is a subnormal L-subgroup of λ where $\lambda \in L(\mu)$ such that $\eta \subseteq \lambda$ with defect c where $c \leq m$,

ii. Let $\theta \in NL(\mu)$. Then, $\eta \circ \theta$ is a subnormal L-subgroup of μ with defect c where $c \leq m$.

It can be seen easily that the intersection of any finite set of subnormal L-subgroups is again subnormal. More generally:

Theorem 1.114 Let $\{\theta_i : i \in I\}$ be a family of subnormal L-subgroups such that defect of θ_i is m_i where $m_i \leq m$. Then, $\underset{i \in I}{\cap} \theta_i$ is a subnormal L-subgroup of μ with defect c where $c \leq m$.

Our next result determines the transitivity of the notion of subnormality.

Theorem 1.115 Let $\eta, \theta \in L(\mu)$ such that η is a subnormal L-subgroup of θ with defect m and θ is a subnormal L-subgroup of μ with defect n. Then, η is a subnormal L-subgroup of μ with defect $m+n$.

The following theorem establishes that the subnormality in L-setting is also preserved under the action of a homomorphism and its inverse image:

Theorem 1.116 Let $\eta \in L(\mu)$ and $f : G \rightarrow K$ be a group homomorphism. Then,

i. if η is a subnormal L-subgroup of μ with defect n, then $f(\eta)$ is a subnormal L-subgroup of $f(\mu)$ with defect m where $m \leq n$,

ii. if η is a subnormal L-subgroup of μ with defect n, then $f^{-1}(\eta)$ is a subnormal L-subgroup of $f^{-1}(\mu)$ with defect m where $m \leq n$, provided that the group homomoirphism f is onto.

6.2 Subnormal L-subgroups and nilpotency

In this subsection, we characterize subnormal L-subgroups by the usual group theoretic subnormality of the level subsets of the given L-subgroups. We shall refer this as a level subset characterization of subnormality. Then, this characterization is used to establish that when the lattice L is an upper well ordered chain, then every L-subgroup of a nilpotent L-group is subnormal. For this purpose, we need to develop a necessary mechanism (see [10, 33]). Here, we present:

Theorem 1.117 Let $\eta \in L(\mu)$ be such that μ and η are jointly supstar. Then,

$$Im\ \eta^{\mu} \subseteq Im\ \left(\mu\eta\mu^{-1}\right) \subseteq Im\mu \cup Im\eta. \tag{66}$$

More generally we have:

Theorem 1.118 Let $\eta \in L(\mu)$ be such that μ and η are jointly supstar. Then, for each i

$$Im\ \eta_{i+1} \subseteq Im\ \left(\eta_i\eta\eta_i^{-1}\right) \subseteq Im\mu \cup Im\eta, \tag{67}$$

where η_i is the *ith* normal closure of η in μ.

Theorem 1.119 Let $\eta \in L(\mu)$ be such that μ and η are jointly supstar. Then for each i,

i. η and η_i are jointly supstar.

ii. if $\eta \in L(\mu)$ and μ and η be jointly supstar, then the L-subset $\eta_i\eta\eta_i^{-1}$ possesses sup-property

Below, we discuss the level subsets and strong level subsets of the normal closure of η in μ.

Theorem 1.120 Let $\eta \in L(\mu)$. Then,

i. $(\eta^{\mu})_a = (\eta_a)^{\mu_a}$ for each $a \leq \eta(e)$, provided μ and η are supstar,

ii. $(\eta^{\mu})_a^{>} = \left(\eta_a^{>}\right)^{\mu_a^{>}}$ for each $a < \eta(e)$, provided L is a chain.

Theorem 1.121 Let $\eta \in L(\mu)$ be such that η and μ are jointly supstar. Then, η is subnormal having defect at most n if and only if each level subset η_a is subnormal having defect atmost n where $a \leq \eta(e)$.

Theorem 1.112 Let G be a group and H be its subgroup. Then, H is a subnormal subgroup of G if and only if 1_H is a subnormal L-subgroup of 1_G.

The strong level subset characterization of subnormal L-subgroup can be obtained easily.

Theorem 1.123 Let L be chain and $\eta \in L(\mu)$. Then, η is subnormal having defect at most n if and only if each strong level subset $\eta_a^>$ is subnormal having defect atmost n where $a < \eta(e)$.

Theorem 1.124 Let L be an upper well ordered chain. Let $\eta \in L(\mu)$ and η be nilpotent having tip a_0 and tail t_0 and $a_0 \neq t_0$. If $\theta \in L(\eta)$ and has the tail t_0, then θ is a subnormal L-subgroup of η.

In a forthcoming paper [10], we develop a mechanism in order to tackle the join problem for subnormal L-subgroup and we prove that:

Theorem 1.125 Let η and θ be subnormal L-subgroups of μ. Let $\eta(e) = \theta(e)$ and $\eta \circ \theta \in L(\mu)$. Then, the following are equivalent:

i. $\eta \circ \theta$ is subnormal in μ,

ii. η^θ is subnormal in μ,

iii. $[\eta, \theta]$ is subnormal in μ.

7. Discussion and analysis

The subject matter discussed in this work presents a systematic and compatible theory of L-subgroups(fuzzy subgroups). In this work, firstly we replace ordinary fuzzy subsets by lattice valued fuzzy subsets. This puts our work beyond the purview of Tom Head's metathoerem [35]. This is contrary to the situation of fuzzy group theory, where the results which are obtained, become simple instances of application of metatheorem. Secondly, throughout our work, we have replaced the parent structure of an ordinary group by an L-subgroup which is called an L-group. In order to carry out our studies successfully, we need to use the notion of normality of a fuzzy subgroup in a fuzzy group defined by Wu [22] rather than the notion of normality of a fuzzy subgroup in a group introduced by Liu [21]. Without following this approach, we can not construct various types of series of L-subgroups, such as the subinvariant series, normal closure series, subnormal series or derived series: even the results of classical group theory such as a characteristic subgroup of a normal subgroup of a group is a normal subgroup of the given group or the property of transitivity of characteristic subgroup can not be formulated and extended to the L-setting. By following Wu's normality, we have successfully extended the above results to the L-setting. Moreover, we have proved that the commutator L-subgroup of an L-group is a characteristic L-subgroup. Furthermore, we proved that a commutator L-subgroup of a supersolvable L-subgroup is nilpotent. Thus an application of Wu's normality establishes a very high degree of compatibility among various notions studied in L-group theory.

Another deviation in our work from the work of earlier researchers in fuzzy group theory is the construction of the trivial L-fuzzy subgroup. This is a proper fuzzification of the notion of trivial subgroup of an ordinary group and it makes possible a successful study of various types of series, arising while dealing with the notions of nilpotency, solvability, supersolvability, subnormality etc.. While defining a trivial L-subgroup, the notion of infimum of an L-set comes into play. So is the case at various other places in our investigations where the infimum of an L-subgroup plays a significant role. This is due to the fact that we are carrying out our investigations in an L-group rather than an ordinary group and we use Wu's normality in place of Liu's normality. There are several peculiarities of L-setting which involves the notion of infimum of L-subgroups. For example Theorem 1.60 states that the infimum of the set product of two L-subsets lies in between the meet and the join of the infimums of given L-subsets. This result has been used in Theorem 1.81 wherein we have obtained a necessary and sufficient condition for the set product of two trivial L-subgroups of an L-group to be a trivial L-subgroup. This result is instrumental in establishing a sufficient condition for the set product of two nilpotent L-subgroups to be nilpotent. Moreover, the role of infimum (tail) of L-subgroups is displayed while dealing with homomorphic and inverse homomorphic images of commutator and solvable L-subgroups. To show that the the concepts of nilpotency, solvability and supersolvability are closed under the formation of subgroups, the notion of infimum is again helpful. The infimums (tails) of all the members of almost all the series discussed in our work are identical with the tails of their respective trivial L-subgroups.

Last but not the least, another pleasing feature of our study is the formation of lattices of normal L-subgroups and characteristic L-subgroups of an L-group. In the process, the notion of sup-property has played a very significant role. We obtain a characterization of sup-property by using the notion of image set of the given L-subset. This characterization of sup-property forms the basis for our generalization. This gave rise to the notions of supstar family of L-subsets and jointly supstar L-subsets. The notion of image of an L-subset is intimately related with the above mentioned concepts. Theorem 1.36 shows that each member of supstar family possesses sup-property. Further, in Theorem 1.70, a relationship has been established between the image of ith members of descending central series of an L-subgroup and the image of the given L-subgroup. A similar relationship has been obtained in Theorem 1.84 and Theorem 1.120.

These concepts are subsequently used in the development of L-group theory. A co-ordinated approach between all these concepts paved a way for a successful development of L-group theory.

8. Conclusion

It has been said earlier that L-group theory developed in [10, 25–34] is a very rich generalization of classical group theory. For if we replace the lattice L by the closed unit interval L, then we retrieve fuzzy group theory. Moreover, if we replace L by the two elements set $\{0,1\}$, we obtain results of classical group theory as simple corollaries of corresponding results of L-group theory.

In view of this development we suggest the researchers, working in the areas of other fuzzy algebraic structures to shift their studies to lattice valued fuzzy sets (L-subsets). Also, in order to obtain consistency, the parent structure of classical algebra should be replaced by the corresponding fuzzy algebraic structures.

For those who are involved in active research in these areas, we propose here few research problems: The formation of quotient structure in fuzzy algebraic structures has been a problem child since its very inception. Its proper formulation is still awaited. We invite the researchers to construct a quotient of L-group μ by a normal L-subgroup in μ in the sense of Wu [22]. The second problem which is likely to be tackled more easily is related to nilpotent L-subgroups of an L-group which is discussed in the present work, that is, the investigation of nilpotency by upper central series. The upper central series is not yet formulated in the theory of L-subgroups.

Finally, to mention the further richness of this generalization, we emphasize that here we study the group theoretic properties of posets of subgroups of a group or in particular chains of subgroups of a group rather than properties of a single subgroup. This way, L-group theory provides us a new language and a new tool for the study of the classical group theory. The classical group theory has been founded on abstract sets and therefore the language used for its development is formal set theory. On the other hand, L-group theory expresses itself through the language of functions. The functions which are lattice valued. Therefore the approach adopted in the studies of L-group theory can be looked upon as a modernization of the approach of classical group theory.

Dedication

I dedicate this work to Prof. Naseem Ajmal, one of the greatest fuzzy algebraists and my Ph. D. supervisor.

Author details

Iffat Jahan
Department of Mathematics, Ramjas College, University of Delhi, India

*Address all correspondence to: ij.umar@yahoo.com

References

[1] Rosenfeld A. Fuzzy groups. Journal of Mathematical Analysis and Applications. 1971;**35**:512-517. DOI: 10.1016/0022-247X(71)90199-5

[2] Abu-Osman MT. On the direct product of fuzzy subgroups. Fuzzy Sets and Systems. 1984;**12**:87-91. DOI: 10.1016/0165-0114(84)90052-6

[3] Akgul M. Some properties of fuzzy subgroups. Journal of Mathematics. 1988;**133**:93-100. DOI: 10.1016/0022-247X(88)90367-8

[4] Anthony JM, Sherwood H. A characterisation of fuzzy subgroups. Fuzzy Sets and Systems. 1982;**7**:297-305. DOI: 10.1016/0165-0114(82)90057-4

[5] Asaad M, Zaid AS. On contribution to the theory of fuzzy subgroups. Fuzzy Sets and Systems. 1996;**77**:355-369. DOI: 10.1016/0165-0114(95)00046-1

[6] Das PS. Fuzzy groups and level subgroups. Journal of Mathematical Analysis and Applications. 1981;**84**:264-269. DOI: 10.1016/0022-247X(81)90164-5

[7] Goguen JA. L-sets. Journal of Mathematical Analysis and Applications. 1967;**18**:145-174. DOI: 10.1016/0022-247X(67)90189-8

[8] Gupta KC, Sarma BK. Commutator fuzzy groups. The Journal of Fuzzy Mathematics. 1996;**4**:655-663

[9] Gupta KC, Sarma BK. Nilpotent fuzzy groups. Fuzzy Sets and Systems. 1999;**101**:67-176. DOI: 10.1016/S0165-0114(97)00067-5

[10] Jahan I, Ajmal N. The join problem of Subnormal L-subgroups of an L-group

[11] Kim JG. Fuzzy orders relative to fuzzy subgroups. Information Sciences. 1994;**80**:341-348. DOI: 10.1016/0020-0255(94)90084-1

[12] Kuroki N. Fuzzy congruences and fuzzy normal subgroups. Information Sciences. 1991;**60**:247-259. DOI: 10.1016/0020-0255(92)90013-X

[13] Kim JG. Commutative fuzzy sets and nilpotent fuzzy groups. Information Sciences. 1995;**83**:161-174. DOI: 10.1016/0020-0255(94)90084-1

[14] Kumar R. Fuzzy characteristic subgroup of a group. Fuzzy Sets and Systems. 1992;**48**:397-398. DOI: 10.1016/0165-0114(92)90356-9

[15] Kumar IJ, Saxena PK, Yadav P. Fuzzy normal subgroups and fuzzy quotients. Fuzzy Sets and Systems. 1992;**46**:121-132. DOI: 10.1016/0165-0114(92)90273-7

[16] Makamba BB. Direct product and isomorphism of fuzzy subgroups. Information Sciences. 1992;**65**:33-43. DOI: 10.1016/0020-0255(92)90076-K

[17] Malik DS, Mordeson JN. Fuzzy Subgroups Of Abelian Groups. Chinese Journal of Mathematics. 1991;**19**:129-145. https://www.jstor.org/stable/43836412

[18] Mashinchi M, Salili S, Zahedi MM. Lattice structures of fuzzy subgroups. Iranian Mathematical Society. 1992;**18**:17-29

[19] Ovchinnikov S. On the image of L-fuzzy group. Fuzzy Sets and Systems. 1998;**94**:129-131. DOI: 10.1016/S0165-0114(96)00361-2

[20] Sarma BK, Ali T. Weak and strong fuzzy homomorphisms of Groups. The

Journal of Fuzzy Mathematics. 2004;**12**: 357-368

[21] Liu WJ. Fuzzy invariant subgroups and fuzzy ideals. Fuzzy Sets and Systems. 1982;**21**:133-139. DOI: 10.1016/0165-0114(82)90003-3

[22] Wu W. Normal fuzzy subgroups. Fuzzy Math. 1981;**1**:21-30

[23] Martinez L. Fuzzy subgroups of fuzzy groups and fuzzy ideals of fuzzy rings. Journal of Fuzzy Mathematics. 1995;**3**:833-849

[24] Martinez L. Prime and Primary L-ideals and L-rings. Fuzzy Sets and Systems. 1999;**101**:489-494. DOI: 10.1016/S0165-0114(97)00114-0

[25] Ajmal N, Jahan I. Generated L-subgroup of an L-group. Iranian Journal of Fuzzy Systems. 2015;**12**:129-136. DOI: 10.22111/IJFS.2015.1988

[26] Ajmal N, Jahan I. An L-point characterization of normality and normalizer of an L-subgroup of an L-group. Fuzzy Information and Engineering. 2014;**6**:147-166. DOI: 10.1016/j.fiae.2014.08.002

[27] Ajmal N, Jahan I. A study of normal fuzzy subgroups and characteristic fuzzy subgroups of a fuzzy group. Fuzzy Information and Engineering. 2012;**2**: 123-143. DOI: 10.1007/s12543-012-0106-0

[28] Ajmal N, Jahan I. Nilpotency and theory of L-Subgroups of an L-group. Fuzzy Information and Engineering. 2014;**6**:1-17. DOI: 10.1016/j.fiae.2014.06.001

[29] Ajmal N, Jahan I. Solvable L-subgroup of an L-group. Iranian Journal of fuzzy Systems. 2015;**12**:151-161. DOI: 10.22111/IJFS.2015.2025

[30] Ajmal N, Jahan I. Normal Closure of an L-subgroup of an L-group. The Journal of Fuzzy Mathematics. 2014;**22**: 115-126

[31] Ajmal N, Jahan I, Davvaz B. Nilpotent L-subgroups and the set product of L-subsets. European Journal of Pure and Applied Mathematics. 2017; **10**:255-271. https://www.ejpam.com/index.php/ejpam/article/view/2822

[32] Ajmal N, Jahan I, Davvaz B. Solvability, Supersolvability and Schreier Refinement Theorem for L-subgroups. Fuzzy Information and Engineering. 2021;**13**:470-496. DOI: 10.1080/16168658.2021.1997444

[33] Ajmal N, Jahan I, Davvaz B. Subnormality and Theory of L-subgroups. European Journal of Pure and Applied Mathematics. 2022;**15**: 2086-2115. DOI: 10.29020/nybg.ejpam.v15i4.4548

[34] Jahan I, Davvaz B, Ajmal N. Nilpotent L-subgroups Satisfy the Normalizer Condition. Journal of Intelligent Fuzzy Systems. 2017;**33**: 1841-1854. DOI: 10.3233/JIFS-17334

[35] Head T. A metatheorem for deriving fuzzy theorems from crisp versions. Fuzzy Sets and Systems. 1995;**73**: 349-358. DOI: 10.1016/0165-0114(94)00321-W

[36] Mukherjee NP, Bhattacharya P. Fuzzy normal subgroups and fuzzy cosets. Information Sciences. 1984;**34**: 225-239. DOI: 10.1016/0020-0255(84)90050-1

[37] Mukherjee NP, Bhattacharya P. Fuzzy groups: Some group theoretic analogs. Information Sciences. 1986;**39**: 247-268. DOI: 10.1016/0020-0255(86)90039-3

[38] Mukherjee NP, Bhattacharya P. Fuzzy groups: Some group theoretic analogs II. Information Sciences. 1987; **41**:77-91. DOI: 10.1016/0020-0255(87) 90006-5

[39] Kim DS. Characteristic fuzzy groups. Korean Mathematical Society. 2003;**18**: 21-29

[40] Sidky FI, Mishref MAA. Fully invariant, characteristic and S-fuzzy subgroups. Information Sciences. 1991; **55**:27-33. DOI: 10.1016/0020-0255(91) 90003-D

[41] Zaid AS. On generalized characteristic fuzzy subgroup of finite group. Fuzzy Sets and Systems. 1991;**43**: 247-259. DOI: 10.1016/0165-0114(91) 90080-A

[42] Ray S. Solvable fuzzy groups. Information Sciences. 1993;**75**:47-61. DOI: 10.1016/0020-0255(93)90112-Y

[43] Sarma BK. Solvable fuzzy groups. Fuzzy Sets and Systems. 1996;**106**: 463-467. DOI: 10.1016/S0165-0114(97) 00264-9

[44] Zaid AS. On fuzzy subnormal and pronormal subgroup of a finite group. Fuzzy Sets and Systems. 1992;**47**: 355-369. DOI: 10.1016/0165-0114(92) 90300-S

[45] Gratzer G. General Lattice Theory. 1st ed. New York: Academic Press; 1978

[46] Ajmal N. Fuzzy groups with sup property. Information Sciences. 1996;**93**: 247-264. DOI: 10.1016/0020-0255(96) 00040-0

Section 2

Applications and Implementations of Fuzzy Logic Systems for Humanity

Chapter 5

Fuzzy Photogrammetric Algorithm for City Built Environment Capturing into Urban Augmented Reality Model

Igor Agbossou

Abstract

Cities are increasingly looking to become smarter and more resilient. Also, the use of computer vision takes a considerable place in the panoply of techniques and algorithms necessary for the 3D reconstruction of urban built environments. The models thus obtained make it possible to feed the logic of decision support and urban services thanks to the integration of augmented reality. This chapter describes and uses Fuzzy Cognitive Maps (FCM) as computing framework of visual features matching in augmented urban built environment modeling process. It is a combination of the achievements of the theory of fuzzy subsets and photogrammetry according to an algorithmic approach associated with the ARKit renderer. In this experimental research work, part of which is published in this chapter, the study area was confined to a portion of a housing estate and the data acquisition tools are in the domain of the public. The aim is the deployment of the algorithmic process to capture urban environments built in an augmented reality model and compute visual feature in stereovision within FCM framework. The comparison of the results obtained with our approach to two other well-known ones in the field, denotes the increased precision gain with a scalability factor.

Keywords: fuzzy cognitive maps, fuzzy sets, photogrammetry, urban augmented reality model, fuzzy stereovision matching constraints

1. Introduction

The use of advanced scientific computation methods and techniques is classic in geography, land use and regional planning [1–8]. Indeed, the study and analysis of geographical spaces such as cities for example, which themselves have acquired the qualification of complex system [1, 2, 5, 9–14] are an illustration of this classic usage [15–24]. Among these scientific computational approaches is also fuzzy inference logic [21, 22, 25, 26]. As part of the research work reported in this chapter, we relied on the scientific achievements of fuzzy inference systems [27–30] to integrate into our methodological approach Fuzzy Cognitive Maps (FCM) [31–34] in the process of visual features matching computation. It's applied through the collection of captured photography of urban built environment to build an augmented urban reality model [35–40].

Thereby, thematic analyzes of built urban environments require the acquisition of 3D urban landscape data, street furniture and several other real visual data. The data to be processed are bi-dimensional (2D) images captured from the tri-dimensional (3D) scene. The objects in 3D are generally composed of related parts that joined from the whole object. In computer graphics, we usually use a specialized software, for instance, Maya [41] or Blender [42, 43] to interactively create models or procedural 3D modeling [44–48] which creates a mathematical representation of a 3D object. It is common to use a few photographs as references and textures to generate models using a modeling tool. When it comes to 3D modeling and urban spaces [49–51], the more systematic introduction of photographs as input to generate a photorealistic 3D model of a built environment is called « Image-based Modeling" [52] and can generate models for objects physically existing. More importantly, such a modeling process can be automated, and therefore can be scaled up for applications [52]. More fundamentally, how to recover the lost third dimension of objects from a collection of 2D images is one of the main objectives of computer vision [53] and the technical challenge resolved in this work. Fortunately, the relations in 3D are preserved in 2D [42, 44, 45, 47, 54]. Hence, we can exploit this fact by considering specific and basic elements which are related to other elements in the 2D images. Those specific and basic elements are stereo correspondence features: epipolar [55], similarity [56], smoothness [57], ordering [58] and uniqueness [59].

Indeed, the use of photogrammetry, which is a technique that consists of taking measurements in a scene, using the parallax obtained between images acquired from different points of view, proves to be an excellent approach for producing captures that conform to the reality [53]. To better manage the parallax during the 3D reconstruction, we combined fuzzy classification algorithm [8, 60, 61] to the photogrammetric processing within the framework of the well-established soft computing technique called Fuzzy Cognitive Map (FCM) [62–67]. Our Fuzzy Photogrammetric Algorithm Kernel (FPAK) applied to 3D reconstruction from images precisely becomes the meeting point of computer graphics and vision, with the finalized 3D representation of urban built environment.

The rest of the chapter is organized as follows: Section 2 presents background of FCM and its mathematical formalization we adopted [22]. Section 3 expose the core of this chapter: materials and methods. Section 4 presents with the experimental results obtained. The conclusion of the chapter puts lights on the future.

2. FCM background

Well-developed modeling methodology for complex systems that allow to describe the behavior of a system in terms of concepts, Fuzzy Cognitive Maps (FCM) are powerful tools for modeling dynamic systems that was introduced by Kosko [32, 68, 69]. The resulting model describes expert knowledge (semantic concepts and/or computed values for example) of complex systems with high dimensions and a variety of factors. The scientific community is expressing a growing interest about the theory and application of FCMs in complex systems, and their validity and usefulness has been proved in various fields [22, 62–67, 70, 71]. FCMs are fuzzy causality backpropagation approach of modeling which combine fuzzy logic, nonlinear computing, semantic and neural networks.

2.1 Theoretical foundation of FCM

FCMs are fuzzy signed directed graphs with feedback. They are appropriate to encode knowledge thanks to concepts organized and causally linked to each other with weightings. Each concept is materialized by a network node. Different FCM networks have been used as a decision modeling tool under different approaches [63–67, 72]. FCMs are based on the theory of fuzzy logic and fuzzy subsets, thus improving the ability of cognitive maps to present and model qualitatively and quantitatively dynamic nonlinear systems. So, FCM is a soft computing modeling technique used for dynamic causal knowledge acquisition and process reasoning. Under its most general approach, each concept represents an entity, a state, a variable, or a feature of the system. An FCM embeds the topology of a fuzzy signed direct graph and a nonlinear neural networks feedback dynamic [26, 33, 61]. Concepts are equivalent to neurons which state value is not binary but belongs to a fuzzy subset. The value w_{ij} of the directed edge from causal concept C_i to concept C_j measures how much C_i causes C_j. Value w_{ij} belongs the fuzzy causal interval $[-1, +1]$, $w_{ij} = 0$ indicates no causality; $w_{ij} > 0$ indicates causal increase, this means that C_j increases as C_i increases and vice versa, C_j decreases as C_i decreases; $w_{ij} < 0$ indicates causal decrease or negative causality. C_j decreases as C_i increases and C_j increases as C_i decreases.

Depending on the direction and size of this effect, and on the threshold levels of the dependent concepts, the affected concepts may subsequently change their state as well, thus activating further concepts within the network. Because FCMs allow feedback loops, newly activated concepts can influence concepts that have already been activated before. As a result, the activation spreads in a nonlinear fashion through the FCM net until the system reaches a stable limit cycle or fixed point.

2.2 FCM representation

To illustrate the description made above of FCMs, we will consider one, composed of 5 concepts and 10 causality links in total as shown in **Figure 1**. Concepts variables are represented by nodes, such as C_1, C_2, C_3, C_4 and C_5.

In the relation $C_1 \rightarrow C_2$, C_1 is said to impact C_2. So, C_1 is the causal variable, whereas C_2 is the effect variable, and the intensity of the causality is expressed by the value of w_{12}. Also, in the relation $C_2 \rightarrow C_1$, C_2 is said to impact C_1, and the intensity of the causality is expressed by the value of w_{21}. Each concept is characterized by a

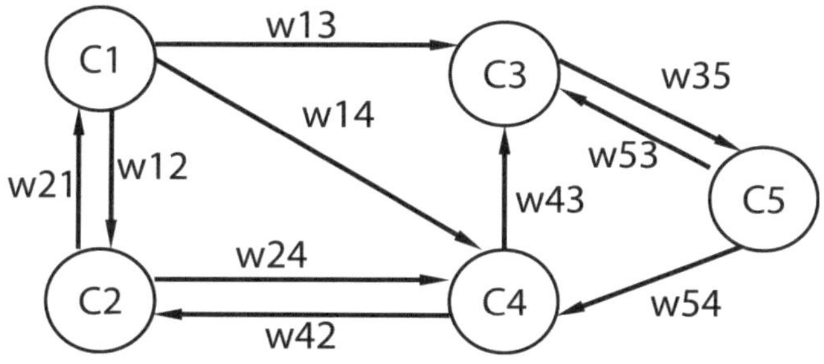

Figure 1.
Simple fuzzy cognitive map model illustration.

number A_i that results from its computed value through the transformation of the real value of the hole system's variables.

There are 3 possible types of causal relationships between concepts:

- $w_{ij} > 0$ which indicates positive causality between two concepts.
- $w_{ij} < 0$ which indicates negative causality between two concepts.
- $w_{ij} = 0$ which indicates the absence of causality between two concepts.

The value of w_{ij} indicates how strongly concept C_i influence concept C_j. The sign of w_{ij} indicates whether the relationships between concept C_i and C_j is direct or inverse. The direction of causality indicates whether concept C_i causes concept C_j or vice versa. These parameters must be considered when a value is assigned to weight w_{ij}.

2.3 Mathematical formalization of FCM

The operation of FCMs is based on an inferential process whose dynamics can be formalized mathematically. A FCM model acts as a network of threshold or continuous concepts [64, 66, 68, 69]. At this level, they differ from a simple neural network because they are based on extracting knowledge from experts [33, 64, 73] and do not require a data input layer. The nonlinear structure of each concept is expressed during the dynamics of the whole system through backpropagation [74, 75]. Then, the value A_i^{t+1} for each concept C_i at each time step is calculated [22, 65, 74] by the following general rule:

$$A_i^{t+1} = f\left(k_1 \sum_{\substack{j=1 \\ j \neq i}}^{n} W_{ji} A_j^t + k_2 A_i^t \right) \tag{1}$$

The k_1 expresses the influence of the interconnected concepts in the configuration of the new value of the concept A_i and k_2 represents the proportion of the contribution of the previous value of concept in the computation of the new value. This formulation assumes that a concept links to itself with a weight $w_{ii} = k_2$. Namely, A_i^t and A_i^{t+1} are respectively the values of concept C_i at times t respectively $t + 1$. w_{ji} is the weight of the interconnection from concept C_j to concept C_i and f is a threshold function defined in Eq. (2). The unipolar sigmoid function is the most used threshold function [57, 65, 67] where $\lambda > 0$ determines the steepness of the continuous function f. For the purposes of this research, the value of λ is fixed at unity, i.e. 1. The sigmoid function ensures that the calculated value of each concept will belong to the interval [0,1].

$$f(x) = \frac{1}{1 + e^{-\lambda x}} \tag{2}$$

3. Materials and methods

Physical based rendering 3D simulation of large-scale urban built environment processes is one of greatest challenges of modern computing techniques in urban studies and regional planning [17, 39, 47, 50]. In fact, urban systems are naturally complex by own [1–5, 14]. Simulation allows us to understand the reasons and effects of events and situations in a real system. Moreover, it allows us to predict the results of actions on future states of the system. The level of detail [15, 39, 45, 50] between simulation results and real system behavior depends on the model used. More-precise models with large data may reflect reality in a more-precise manner; however, the complexity directly influences the time required to compute model changes.

3.1 Urban study area

The model created in this study covers a portion of a new housing estate under construction in the town of Belfort in France. The area of the development project for building individual houses has 25 plots of 600 to 900 m^2. **Figure 2** provides an overview of the area called "Jardins du MONT".

Indeed, it is a real estate program whose architecture of the houses is contemporary, of high quality and of superior range located less than 10 minutes by car, bus, or bike from the city center of Belfort. We are also less than 10 minutes' walk from the heart of the "Techn'Hom" business park (GE, Alstom...), one of the economic lungs of the town, with an exceptional view of Belfort, its fortifications and the surroundings, all integrated in a green, calm and privileged urban setting.

The general framework of our research work includes 3D spatial analysis, the temporal evolution of new housing estates and the deployment of smart cities, with scientific tools in artificial intelligence. Also, it seemed legitimate to us to take an interest in this portion of the city under construction to experiment with our approach which is the subject of this chapter: create an augmented reality scene model of the built environment through the combination of photogrammetry [76–81] and fuzzy modeling techniques.

3.2 Sensor for data acquisition

In addition to the question of costs, the spatial scale of the data to be collected as well as the expected quality dictate the choice of tools to be preferred. There are

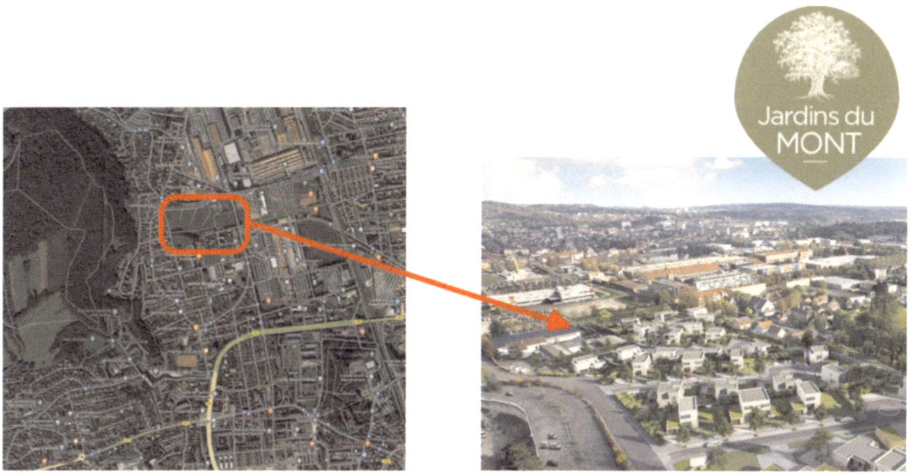

Figure 2.
Urban study area "Jardins du MONT", Belfort (France).

several tools for Geodata collection [38, 49]: Total Stations, Global Navigation Satellite System (GNSS) receivers, Light Detection and Ranging (LiDAR) scanner, Static Terrestrial Laser Scanning (STLS), Airborne Laser Scanning (ALS), Helicopter Laser Scanning (HLS), Mobile Laser Scanning (MLS), Drone, Tablets and Smartphones. As part of this experimental study, we use portable and mobile sensor which smartphones are equipped with. And for good reason, the sensors of these modern devices perfectly meet the requirements relating to the acquisition of data for photogrammetry [82]. Range (or depth) data is crucial for understanding and working with the 3D scene projected onto a 2D plane forming an image. There are multiple ways to obtain such information [83–87], either using a depth sensor or estimating depth. A depth sensor is a device that provides the distance from the sensor to an element in the scene, although it is possible to collect distance information using two or more RGB cameras from a scene.

Due to its following features: wide color capture for photos and live photos, lens correction, retina flash, auto image stabilization, burst mode, etc. we used the iPhone 13 Pro Max as a sensor for acquiring images to feed the model. **Figure 3** illustrates it.

3.3 Data collection principles and quality requirements

When capturing images for augmented reality, we use a large part of the image sensor. To be more precise, it's an area of 3840 × 2880 pixels on the iPhone 13 Pro. Then, we use a process called binning [88, 89]. It works as follows: Binning takes a region of 2×2 pixels, averages the pixel values, and writes back a single pixel. This has two significant advantages. First, image dimensions are reduced by a factor of two, in this case, it downscales to 1920 × 1440 pixels. As a result of this, each frame consumes way less memory and processing power. This allows the device to run the camera at up to 60 frames per second and frees up resources for rendering. Secondly, this process offers an advantage in low light environments, where the averaging of pixel values reduces the effects of sensor noise.

Images captured by a camera are geometrically warped by small imperfections in the lens. To project from the 2D image plane back into the 3D world, the images must be distortion corrected, or made rectilinear. Lens distortion is modeled using a one-dimensional lookup table of 32-bit float values evenly distributed along a radius from the center of the distortion to a corner, with each value representing a magnification of the radius. This model assumes symmetrical lens distortion [88].

Capturing scenes with iPhone is a computer vision technology that one can leverage to easily turn images of real-world objects into detailed 3D models. We begin by

Figure 3.
iPhone 13 pro max used as sensor for data acquisition.

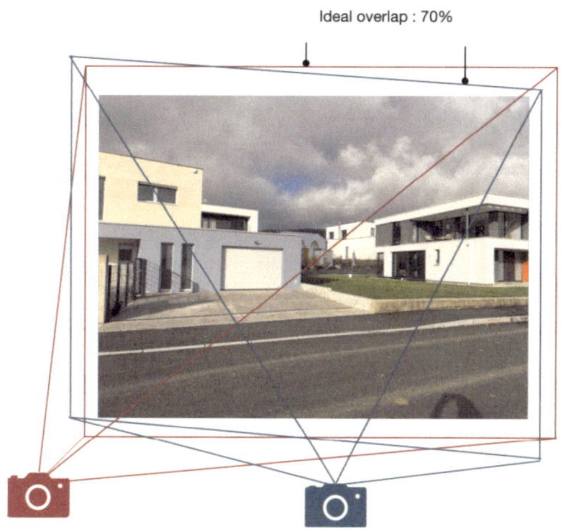

Figure 4.
Ideal overlap to respect when capturing built environment.

taking photos of the urban built environment from various angles with an iPhone. To photograph all the area with the ability to match landmarks between images we must move the camera around, taking photographs from different angles at different heights.

To ensure landmarks matching between overlapping photographs, camera settings must be consistent as possible from shot to shot. **Figure 5** illustrates a sample of captured data. The reading direction of the photos is indicated there: start-end.

The number of pictures need to create an accurate 3D representation varies depending on the quality of the pairs of photographs making up the sequences in the collection, the complexity and size of the built environment. In addition, adjacent shots must have substantial overlap. So, we position sequential images, so they have a 70% overlap or more ($0.7 \leq$ overlap ≤ 0.9) as illustrated in **Figure 4**. Anything less than 50% overlap between neighboring shots, and the 3D reconstruction process may fail or result in a low-quality augmented reality model [15, 52]. Doing an aperture setting narrow enough to maintain a crisp focus is recommended [53, 58]. The spatial precision between the pairs of images and the chromatic density of the textures are a guarantee of the quality of the images collected for the 3D reconstruction of built urban environments. Accordingly, key factors ensuring good quality of input data [15, 52, 53, 58, 90] are summarized in **Table 1**.

Our photographic database is made up of 800 photos taken in compliance with the overlap constraints to feed the model. The entire collection is organized into 799 image pairs. A first step consists in sorting the truly calibrated image pairs according to the constrained constraints of the stereovision image matching.

3.4 Image matching in stereovision within FCM framework

The image matching in stereovision [89, 91–94] is the process of identifying the corresponding points in two images that are cast by the same physical point in the tri-dimensional space. This can be carried out pixel by pixel or identifying significant features in the images, such as edges, regions or interest points.

Hence, the stereo correspondence problem can be defined in terms of finding pairs of true matches, namely, pairs of edge segments in two images that are generated by

Figure 5.
Sample of captured urban built environment dataset

Factor	Description	Fuzzy threshold value
Range or depth	Distance between camera and scene	Low
Sensor quality	The resolution of de sensor	High
Overlap	Superposition rate between two consecutive photographs	$0.7 \leq$ overlap ≤ 0.9
Image texture	Texture and texture variance	High

Table 1.
Key factors affecting photogrammetric input images quality.

the same physical edge segment in space. These true matches generally satisfy some constraints:

1. epipolar, given two segments one in the left image and a second in the right one, if we slide one of them along a horizontal direction, i.e. parallel to the epipolar line, they would intersect (overlap) (**Figure 4**);

2. similarity, matched edge segments have similar properties or attributes;

3. smoothness, disparity values in a given neighborhood change smoothly, except at a few depth discontinuities;

4. ordering, the relative position among two edge-segments in the left image is preserved in the right one for the corresponding matches;

5. uniqueness, each edge-segment in one image should be matched to a unique edge-segment in the other image.

A large parallax factor value causes the background to move more slowly compared to the foreground. A small value makes the foreground and background move at a similar pace. The parallax effect becomes more apparent as the value of parallax factor increases.

According to FCM framework, causal concepts and their activation levels, the system receives as inputs a pair of stereo images left, I_l and right I_r. This pair is processed to extract edge segments and their attributes; each pair of extracted features vectors $(\underset{I_l}{\rightarrow}, \underset{I_r}{\rightarrow})$ is to be matched, the vectors $\underset{I_l}{\rightarrow}$ and $\underset{I_r}{\rightarrow}$ come from I_l and I_r respectively. For each pair $(\underset{I_l}{\rightarrow}, \underset{I_r}{\rightarrow})$ the attribute difference vector $\underset{x}{\rightarrow}$ is computed. In this approach, a pair of edge attributes $(\underset{I_l}{\rightarrow}, \underset{I_r}{\rightarrow})$ defines a causal fuzzy concept C_i,. The Eq. (1) is applied and the initial activation level at the iteration t = 0 is derived from $\underset{x}{\rightarrow}$ as follows in Eq. (3):

$$A_i^0 = \frac{1}{1 + \|\underset{x}{\rightarrow}\|} \tag{3}$$

where $\|\underset{x}{\rightarrow}\|$ is defined as the Euclidean norm. This implies the application of the similarity Gestalt's principle. Hence, our FCM structure is built with as many concepts as pairs of edge attributes, from L_i and L_r, are available. The algorithm is synthesized as follows in **Table 2**.

The correspondence results within the pairs for each of the characteristics are recorded in **Table 3** in number of pairs according to the number of iterations:

It is noted that from iteration n°20 the results remain stable. To test the consolidation of these, we have pushed the number of iterations to 35 without any disruption of stability.

In view of these results of this correspondence calculation phase, only the 744 pairs of photographs respecting the five constraints (epipolar, similarity, smoothness ordering and uniqueness) have been selected to now feed the scene of the augmented urban reality model.

4. Experimental results

We present in this section, the first significant results of the construction of an Urban Augmented Reality (UAR) scene model resulting from the combination of photogrammetry and fuzzy modeling techniques for future analyses. In our

1. Initialization	**Load each concept with its activation level $A_i^{t=0}$ through the Eq. (3);** **Set:** **δ = 0.05, the minimum value of change in classification approach** **t_{max} = 50, maximum of iterations** **α = 0.9, which is the limit indicator of concepts looping in the q network.** **nc: *integer*, the number of concepts from a total of q representing pairs of edge attributes that change their activation level at each iteration. The activation mechanism is that defined in Eq. (1)**

2. FCM process	t = 0
	while (t < t_{max} and nc/q < α) {
	$t \leftarrow t+1; nc \leftarrow 0$
	for (each concept C_i) {
	update A_i^t according to Eq. (1)
	if $(\lvert A_i^{t+1} - A_i^t \rvert > \alpha)$ {
	$nc \leftarrow nc + 1$
	}
	}
	}

3. Output	The activation levels A_i^t for all concepts updated.

Table 2.
Process of image matching in stereovision within FCM framework.

Number of iterations	Epipolar	Similarity	Smoothness	Ordering	Uniqueness
10	788	791	685	778	746
15	790	792	734	783	767
20	791	796	744	785	779
25	791	796	744	785	779
30+	791	796	744	785	779

Table 3.
Image matching in stereovision within FCM framework results.

incremental validation process, we rely on two major works [64, 87, 93, 95] to assess the performance of our method and its robustness for large-scale deployment.

4.1 The urban augmented reality model scene

Based on the 744 image pairs, the total number of photographs therefore amounts to 745. The prodigiously increased computing capacities of mobile devices open opportunities for augmented reality applications. The FPAK we developed a enables the conversion of urban built environment photos into Urban Augmented Reality (UAR) model as illustrated with **Figure 6**. To achieve AUR, we use the kernel in conjunction with Apple ARKit [96] and RealityKit [97] frameworks. The use of RealityKit framework let implement high-performance 3D simulation and rendering. It leverages information provided by the ARKit framework to seamlessly integrate virtual urban built environment into the real world. In turn, the kernel mainly focuses on considering the imprecision of blind spots inherent in the overlapping of shots during the acquisition of photographs to be used as raw materials for the work of

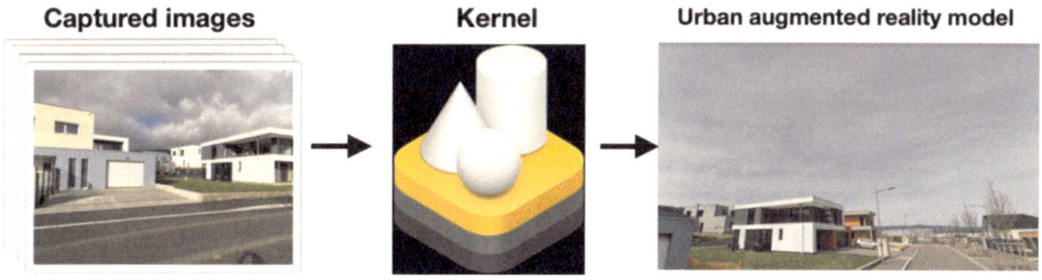

Figure 6.
Experimental UAR model for 3D spatial analysis.

implementing augmented reality scenes. In addition, it provides a flexible architecture that fosters the development augmented reality applications about research in theoretical and quantitative geography like UAR.

4.2 Datasets and input quality analytics

The quality of data (accuracy, precision, and resolution) taken by sensors as smartphones is determined by many factors related to both the capture technique and the physical environment. Ideal physical conditions should favor diffused and homogeneous lighting and all protruding urban objects should have enough space around them. In addition, when taking photos, special attention should be paid to the following object/environment characteristics: sufficient texture detail and minimal reflective surfaces.

To select from the entire set of photographic data, the images meeting these criteria as well as those set out in **Table 1**, a valuation of the threshold values (high, and low) based on a fuzzy set as shown in **Figure 7**.

To ensure the quality of the input data, the input image quality sorting process consisted of sifting through the 1028 raw images captured for the entire study area. Indeed, the 800 photos organized in 799 pairs to constitute the input database are the result of the application of this cleaning process. Also, although variations in the quality of photogrammetric data are attributable to factors beyond the control of the operator, several steps can be taken to increase the likelihood that the data collected will achieve the desired quality. The following three points of vigilance are in order: 1) Consider the expected data collection conditions (e.g., weather, lighting), the quality of the camera and the lens. 2) Using the target range and camera specifications, calculate the desired spacing between successive frames to ensure adequate overlap. The interval [0.7–0.9] is the optimum since the value of 0.7 already gives excellent results. 3) After data collection, review and remove any poor quality/blurry images by manual or automatic means.

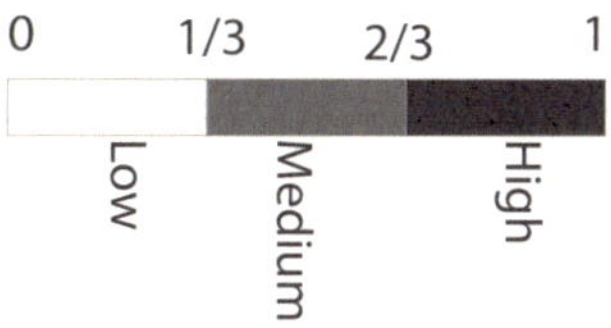

Figure 7.
Fuzzy set assigned to input image quality factor.

4.3 Performance analytics and originality

To evaluate and measure the performance of our FPAK approach associated with the ARKit rendering engine, the results obtained are compared with two other approaches [64, 87, 93, 95] on the same basis of the five constraints referenced in **Table 2**.

The first comparison model is the Deterministic Simulated Annealing (DSA) metaheuristics optimization algorithm. In Pajares and Cruz [95], this strategy for stereovision matching was exploited with satisfactory results. It is a comprehensive approach belonging to the category of methods that incorporate explicit smoothing assumptions and determine all disparities simultaneously by applying a energy minimization process. The limits of this approach are felt when the input database exceeds 82 pairs of stereo images and whose convergence is only reached after 30 iterations [87].

The second comparison model is based on the so-called relaxation labeling approach (RELB). This is a technique proposed by Rosenfeld et al. [98] to account for uncertainty in sensory data interpretation systems and to find the best matches. It uses contextual information as an aid to the classification of a set of interrelated objects by allowing interactions between possible classifications of related objects. In the stereovision paradigm, the problem is to assign unique labels (or matches) to a set of features in an image from a given list of possible matches.

The objective is to assign to each feature (edge segment) a value corresponding to the disparity in a way consistent with certain predefined constraints according to probabilities assigned to the five constraints in the studies [64, 93]. Here, the maximum number of input image pairs is increased to 90 for convergence from the 35th iteration. The results of performance comparison are synthetized in **Table 4**.

Although pioneering works [64, 87, 93, 95] have paved the way for the fuzzy modeling of the constraints inherent in image matching in stereovision applications, the originality of our work is assessed at three distinct levels. First, our method fits perfectly with a professional rendering engine such as ARKit. Second, the five constraints are modeled as concepts within the framework of FCMs. And third, the calculations did not require additional models as in the case of the DSA or RELB based approach. In doing so, the entire modeling chain constituted a fuzzy inference system.

5. Conclusion and future directions

Computer-vision-based API (Application Programming Interface) such as ARKit enable landscape and urban physical feature capture on mobile devices like iPhone with a physically based rendering. They open new possibilities for applications, such as Virtual Geographic Environment (VGE) modeling for 3D spatial analysis. In this chapter, we explored one process of capturing urban built environment into an Urban Augmented Reality Model (UARM) and urban layouts according to the well-established soft computing framework Fuzzy Cognitive Map (FCM). It's a novel application of FCM which let us verify the performance and the robustness of our approach as compared to other existing methods.

Moreover, visualization of the urban development plan using UAR model gives one of the best augmented spatial models for urban planning simulation and 3D spatial analysis. In fact, the paradigm of augmented reality simplifies the process of project

Algorithm	Stereovision matching constraints for UAR modeling						
	Epipolar	**Similarity**	**Smoothness**	**Ordering**	**Uniqueness**	**Maxi pairs of stereo images**	**Iterations need for Convergence**
FPAK	Mapped as coefficients aggregated in the causal weight between concepts	Simple difference vector	Mapped as coefficients aggregated in the causal weight between concepts	Mapped as coefficients aggregated in the causal weight between concepts	Applied by selecting the highest causal concept values	799+	20
DSA	Mapped as an energy minimized by Simulated Annealing	Support Vector Machines	Mapped as an energy minimized by Simulated Annealing	Mapped as an energy minimized by Simulated Annealing	Applied by selecting the highest state value	82	30
RELB	Mapped under the overlapping concept	Bayes probability density estimation	Probabilistic relaxation	Probabilistic relaxation	Applied by selecting the highest probabilities	90	35

Table 4.
Synoptic performance comparison of FPAK with DSA and RELB.

planning, measurement computations, design updates, collection of on-site architecture environment, safety training, etc.

Although UAR model uses multiple tools, it is the best visual aid to get walkthroughs for analyzing the virtual urban development plans. There are specific issues like high computational complexities, networking requirements and storage complexities to be considered. However, in practice, the limitations regarding technical issues can be overcome (to possible extents) as a scope for future research. The proposed method can further enhance the level of understanding of urban built environment by incorporating cloud computing services. We could realize uploading as well as synchronization of information contained in connected devices which feed smart cities.

Thus, the Architecture, Engineering, Construction, and Facility Management (AEC/FM) designs and construction site 3D visuals can be accessible, examinable, and modifiable from any location, irrespective of the location.

Users from different locations can collaborate with each other by accessing these cloud UARM services. The incorporation of cloud UARM for BIM's (Building Information Modeling) 3D visualization of construction layouts does elicit further investigation.

The performance assessment is still in progress. So, for detecting a possible bias of over- and underestimation of the five concepts of image matching due to ARKit, we are investigating two metrics: Mean Absolute Error (MAE) and the non-parametric Spearman's Rank Correlation Coefficient (SRCC).

Author details

Igor Agbossou
Laboratoire ThéMA UMR 6049, Institut Universitaire de Technologie Nord Franche-Comté, Université de Franche-Comté, Belfort, France

*Address all correspondence to: igor.agbossou@univ-fcomte.fr

References

[1] Batty M. Cities and Complexity. Cambridge: MIT Press; 2005

[2] Batty M, Torrens P. Modelling and prediction in a complex world. Futures. 2005;**37**:745-766

[3] Benenson I, Torrens P. Geosimulation: Automata-Based Modelling of Urban Phenomena. Chichester: Wiley; 2002

[4] Berrou JL, Beecham J, Quaglia P, Kagarlis MA, Gerodimos A. Calibration and validation of the legion simulation model using empirical data. In: Aldau WN, Gattermann P, Knoflacher H, Schreckenberg M, editors. Pedestrian and Evacuation Dynamics. New York: Springer Verlag; 2007. pp. 155-166

[5] Portugali J. Self-Organization and the City. New York: Springer-Verlag; 2000

[6] Ioannides YM, Zabel JE. Interactions, neighborhood selection and housing demand. Journal of Urban Economics. 2008;**63**:229-252

[7] Couclelis H. The certainty of uncertainty: GIS and the limits of geographic knowledge. Transactions in GIS. 2003;7:165-175

[8] Biswajeet P, editor. Spatial Modeling and Assessment of Urban Form Analysis of Urban Growth: From Sprawl to Compact Using Geospatial Data. Switzerland: Springer; 2017. p. 331. DOI: 10.1007/978-3-319-54217-1

[9] Dicken P, Lloyd PE. Location in Space: Theoretical Perspectives in Economic Geography. New York: Harper and Row; 1990

[10] Goodchild M. GIScience ten years after ground truth. Transactions in GIS. 2006;**10**:687-692

[11] Longley PA, Goodchild MF, Maguire DJ, Rhind DW. Geographic Information Systems and Science. 2nd ed. Wiley and Sons: Chichester; 2005

[12] Longley PA, Singelton AD. Social deprivation and digital exclusion in England. In: CASA Working Paper, 145. London: UCL Centre for Applied Spatial Analysis; 2008

[13] Morrissey K, Clarke G, Hynes S, O'Donoghue C. Accessibility modelling. In: Bavaud F, Mager C, editors. Handbook of Theoretical and Quantitative Geography, FGSE, Lausanne, Switzerland: University of Lausanne; 2009. p. 457

[14] Pumain D, Sanders L, Saint-Julien T. Villes et auto-organisation. Paris: Economica; 1989

[15] Anders K-H. Level of detail generation of 3D building groups by aggregation and Typification. International Cartographic Conference. Vol. 2. 2005. Available from: https://citeseerx.ist.psu.edu/document?repid=rep1&type=pdf&doi=1f8df00d63c3008b90c68f32cdf498765c9d776d

[16] Batty M. Visually-Driven Urban Simulations: Exploring Fast and Slow Changes in Residential London, CASA, Working Papers Series. Vol. 164. London: UCL; 2011

[17] Bazzanella L, Caneparo L, Corsico F, Roccasalva G, editors. Future Cities and Regions. Simulation, Scenario and Visioning, Governance and Scales. New York, Heidelberg: Springer; 2011

[18] Bittner T, Donnelly M, Winter S. Ontology and semantic interoperability. In: Prosperi D, Zlatanova S, editors.

Large-Scale 3D Data Integration: Challenges and Opportunities. Boca Raton, FL: CRC Press; 2005

[19] Bucher B, Falquet G, Clementini E, Sester M. Towards a typology of spatial relations and properties for urban applications. Usage, Usability, and Utility of 3D City Models. 2012. p. 11. DOI: 10.1051/3u3d/201202010

[20] Gallagher J, Gill LW, Mcnabola A. Numerical modelling of the passive control of air pollution in asymmetrical urban street canyons using refined mesh discretization schemes. Building and Environment. 2012;**56**:232-240

[21] Raufirad V, Heidari Q, Ghorbani J. Comparing socioeconomic vulnerability index and land cover indices: Application of fuzzy TOPSIS model and geographic information system. Ecological Informatics. 2022;**72**:101917. DOI: 10.1016/j.ecoinf.2022.101917

[22] Agbossou I. Fuzzy cognitive maps-based modeling of residential mobility dynamics: GeoComputation approach. Plurimondi. 2017;**17**:169-190

[23] Agbossou I, Provitolo D, Frankhauser P. Expérimentation par voie informatique de la mobilité réside ntielle, XVème Journées de Rochebrune. In: Rencontres interdisciplinaires sur les systems complexes naturels et artificiels. Rochebrune, Megève, France. CD Rom; 2008. pp. 1-13

[24] Agbossou I. Cerner le contexte spatia l par les voisinages dans les modèles ce llulaires en géographie. In: Rencontres interdisciplinaires sur le contexte dans les systèmes complexes naturels et artificiels, Jan 2010. Megève, France; 2010

[25] Marsala C, Bouchon-Meunier B. Entropies et ensembles flous intuitionnistes. In: LFA 2019 - Rencontres francophones sur la Logique Floue et ses Applications. Alès, France: Cépaduès; 2019. pp. 143-148

[26] Coletti G, Bouchon-Meunier B. Fuzzy Similarity Measures and Measure ment Theory. In: IEEE International Conference on Fuzzy Systems 2019 (FUZZ-IEEE 2019). New Orleans, United States: IEEE; 2019

[27] Abbasi F, Allahviranloo T, Abbasbandy S. A new attitude coupled with fuzzy thinking to fuzzy rings and fields. Journal of Intelligent & Fuzzy Systems. 2015;**29**:851-861

[28] Abbasi F, Abbasbandy S, Nieto JJ. A new and efficient method for elementary fuzzy arithmetic operations on pseudo-geometric fuzzy numbers. Journal of Fuzzy, Set Valued Analysis. 2016;**2**: 156-173

[29] Allahviranloo T, Mikaeilvand N. Non zero solutions of the fully fuzzy linear systems. Journal of Computational and Applied Mathematics. 2011;**10**(2): 271-282

[30] Jetter AJ, Kok K. Fuzzy Cognitive Maps for futures studies. A methodological assessment of concepts and methods, Futures. 2014;**61**:45-57, DOI: 10.1016/j. futures.2014.05.002.

[31] Liu ZQ, Satur R. Contextual fuzzy cognitive map for decision support in geographic information systems. IEEE Transactions on Fuzzy Systems. 1999;**7** (5):495-507

[32] Kosko B. Fuzzy Cognitive Maps. International Journal of Man-Machine Studies. 1986;**24**:65-75

[33] Kosko B. Neural Networks and Fuzzy Systems: A Dynamical Systems

Approach to Machine Intelligence. NJ: Prentice Hall; 1992

[34] Xirogiannis G, Stefanou J, Glykas M. A fuzzy cognitive map approach to support urban design. Expert Systems with Applications. 2004;**26**(2):257-268. DOI: 10.1016/S0957-4174(03)00140-4

[35] Unity Real-Time Development Platform. 3D, 2D VR & AR Engine. Available from: https://unity.com/ [Accessed: November 19, 2021]

[36] blender.org—Homeoftheblenderproject—Freeandopen3Dcreationsoftware, Available from: https://www.blender.org/ [Accessed: September 12, 2021]

[37] Verma JK et al. Advances in Augmented Reality and Virtual Reality, Studies in Computational Intelligence. Springer; 2022. DOI: 10.1007/978-981-16-7220-0_2

[38] Karthikeyan OVGSK, Padmanaban S, editors. Smart Buildings Digitalization. Case Studies on Data Centers and Automation. Abingdon, Oxon, OX14 4RN: CRC Press; 2022. p. 314. DOI: 10.1201/9781003240853

[39] Verma JK, Paul S, editors. Advances in Augmented Reality and Virtual Reality. Singapore: Springer; 2022. p. 312. DOI: 10.1007/978-981-16-7220-0

[40] Quan L. Image-Based Modeling. London: Springer; 2010. DOI: 10.1007/978-14419-6679-7

[41] Kuldip A, Dibyendu G. Nature inspired prototype Design of Collision Avoidance Aircraft System and Design of a pair of wing flaps in Autodesk Maya software. Procedia Computer Science. 2016;**89**:684-689. DOI: 10.1016/j.procs.2016.06.036

[42] Naiman JP. AstroBlend: An astrophysical visualization package for blender. Astronomy and Computing. 2016;**15**:50-60. DOI: 10.1016/j.ascom.2016.02.002

[43] Pelayo P et al. CubeSat landing simulations on small bodies using blender. Advances in Space Research. Volume 70, Issue 3 Elsevier, 2022. DOI: 10.1016/j.asr.2022.07.044

[44] Lars K, Leif K. Interactive modeling by procedural high-level primitives. Computers & Graphics. 2012;**36**(5):376-386. DOI: 10.1016/j.cag.2012.03.028

[45] Johannes E et al. Procedural modeling of architecture with round geometry. Computers & Graphics. 2017; **64**:14-25. DOI: 10.1016/j.cag.2017.01.004

[46] Andrew RW et al. Volumetric procedural models for shape representation. Graphics and Visual Computing. 2021;**4**:200018. DOI: 10.1016/j.gvc.2021.200018

[47] Gustavo A et al. Procedural modeling applied to the 3D city model of bogota: A case study. Virtual Reality & Intelligent Hardware. 2021;**3**(5):423-433. DOI: 10.1016/j.vrih.2021.06.002

[48] Mudit G et al. O-2 | development of a 3D modeling tool for procedural planning of ductal stenting. Journal of the Society for Cardiovascular Angiography & Interventions. 2022; **1**(3):100052. DOI: 10.1016/j.jscai.2022.100052

[49] Biljecki F, Ledoux H, Stoter J. Generating 3D city models without elevation data. Computers, Environment and Urban Systems. 2017;**64**:1-18

[50] Peeters A, Etzion Y. Automated recogni- tion of urban objects for

morphological urban analysis. Computers, Environment and Urban Systems. 2012;**36**(6):573-582

[51] Goetz M, Zipf A. OpenStreetMap in 3D – Detailed insights on the current situation in Germany. In: Proceedings of the AGILE'2012 Inter- National Conference on Geographic Information Science. Avignon: AGILE Digital Editions; 2012. pp. 288-292

[52] Kim T-H et al. Smart city and IoT. Future Generation Computer Systems. 2017;**76**:159-162. DOI: 10.1016/j.future.2017.03.034

[53] Yonghuai L et al., editors. 3D Imaging, Analysis and Applications. Second ed. Switzerland: Springer; 2022. DOI: 10.1007/978-3-030-44070-1

[54] Boonsuk W, Gilbert SB, Kelly JW. The impact of three interfaces for 360- degree video on spatial cognition. In: Conference on Human Factors in Computing Systems — Proceedings. New York, USA: ACM Press; 2012. pp. 2579-2588

[55] Puyun L et al. A linear pushbroom satellite image epipolar resampling method for digital surface model generation. ISPRS Journal of Photogrammetry and Remote Sensing. 2022;**190**:56-68. DOI: 10.1016/j.isprsjprs.2022.05.010

[56] Remya R, Nirmala M. A novel similarity metric for image filtering. Optik. 2022;**271**:169977. DOI: 10.1016/j.ijleo.2022.169977

[57] Tahereh B et al. Edge preserving range image smoothing using hybrid locally kernel-based weighted least square. Applied Soft Computing. 2022;**125**:109234. DOI: 10.1016/j.asoc.2022.109234

[58] Xiang W et al. A novel reversible image data hiding scheme based on pixel value ordering and dynamic pixel block partition. Information Sciences. 2015;**310**:16-35. DOI: 10.1016/j.ins.2015.03.022

[59] Owen Saxton W. The image and diffraction plane problem: uniqueness, Reprinted from Advances in Electronics and Electron Physics, Supplement 10, 1978. In: Hÿtch M, Hawkes PW, editors. Advances in Imaging and Electron Physics. Vol. 214. London: Elsevier; 2020. pp. 87-104. DOI: 10.1016/bs.aiep.2020.04.004

[60] Deepak G, Aditya K, Ashish K, Oscar C, editors. Soft Computing for Data Analytics, Classification Model, and Control. Switzerland AG: Springer; 2022. p. 165. DOI: 10.1007/978-3-030-92026-5

[61] Dadios EP, editor. Fuzzy Logic – Algorithms, Techniques and Implementations. London, UK, London, UK: InTech; 2012

[62] Allahviranloo T, Perfilieva I, Abbasi F. A new attitude coupled with fuzzy thinking for solving fuzzy equations. Soft Computing. 2018;**22**(9):3077-3095

[63] Abbasi F, Allahviranloo T. Computational procedure for solving fuzzy equations. Soft Computing. 2021;**25**:1-15. DOI:10.1007/s00500-020-05330-8

[64] Pajares G, de la Cruz JM. Fuzzy cognitive maps for stereovision matching. Pattern Recognition. 2006;**39**(11):2101-2114. DOI: 10.1016/j.patcog.2006.04.003

[65] Adeleke O, Jen T-C. A FCM-clustered neuro-fuzzy model for estimating the methane fraction of biogas in an industrial-scale bio-digester. Energy Reports. 2022;**8**(Supplement 15):576-584. DOI: 10.1016/j.egyr.2022.10.265

[66] Hosseinpour M, Ghaemi S, Khanmohammadi S, Daneshvar S. A hybrid high-order type-2 FCM improved random forest classification method for breast cancer risk assessment. Applied Mathematics and Computation. 2022; **424**:127038. DOI: 10.1016/j.amc.2022.127038

[67] Senthilkumar N et al. Minimally parametrized segmentation framework with dual metaheuristic optimisation algorithms and FCM for detection of anomalies in MR brain images. Biomedical Signal Processing and Control. 2022;**78**:103866. DOI: 10.1016/j.bspc.2022.103866

[68] Kosko B. Hidden patterns in combined and adaptive knowledge networks. International Journal of Approximate Reasoning. 1988;**2**:377-393

[69] Kosko B. Adaptive inference in fuzzy knowledge networks. In: Dubois D, Prade H, Yager RR, editors. Readings in fuzzy sets for intelligent systems. San Mateo: Morgan Kaufman; 1993

[70] Eden C, Ackermann F, Brown I, Eden C, Ackermann F. Making Strategy: The Journey of Strategic Management. London: SAGE; 2006

[71] Eden C, Ackermann F, Cropper S. The analysis of cause maps. Journal of Management Studies. 2007;**29**:309-324

[72] Axelrod R. Structure of Decision: The Cognitive Maps of Political Elites. Princeton, NJ: Princeton University Press; 1976

[73] Kowalski RM, Leary MR. The Social Psychology of Emotional and Behavioral Problems: Interfaces of Social and Clinical Psychology. États-Unis, American Psychological Association; 1999

[74] Papageorgiou E, Stylios CD, Groumpos PP. Active Hebbian learning algorithm to train fuzzy cognitive maps. International Journal of Approximate Reasoning. 2004;**37**(3):219-249

[75] Papageorgiou E, Stylios CD, Groumpos PP. Fuzzy cognitive map learning based on nonlinear Hebbian Rule. In: Gedeon TD, Fung LCC, editors. AI 2003: Advances in Artificial Intelligence. AI 2003. Lecture Notes in Computer Science. Vol. 2903. Berlin, Heidelberg: Springer; 2003. DOI: 10.1007/978-3-540-24581-0_22

[76] Wang Y, Liqiang Z, Takis Mathiopoulos P, Deng H. A gestalt rules and graph-cut-based simplification framework for urban building models. International Journal of Applied Earth Observation and Geoinformation. 2015; **35**(Part B):247-258. DOI: 10.1016/j.jag.2014.09.012

[77] Fang Y, Zhang X, Yuan F, Imamoglu N, Liu H. Video saliency detection by gestalt theory. Pattern Recognition. 2019;**96**:106987. DOI: 10.1016/j.patcog.2019.106987

[78] Xue T, Owens A, Scharstein D, Goesele M, Szeliski R. Multi-frame stereo matching with edges, planes, and superpixels. Image and Vision Computing. 2019;**91**:103771. DOI: 10.1016/j.imavis.2019.05.006

[79] Szeliski R. Computer Vision: Algorithms and Applications, Texts in Computer Science. London: Springer; 2011. DOI: 10.1007/978-1-84882-935-0

[80] Lopes A, Souza R, Pedrini H. A survey on RGB-D datasets. Computer Vision and Image Understanding. 2022; **222**:103489. DOI: 10.1016/j.cviu.2022.103489

[81] Scharstein D, Briggs AJ. Real-time recognition of self-similar landmarks. Image and Vision Computing. 2001;**19**

(11):763-772. DOI: 10.1016/S0262-8856(00)00105-0

[82] Cherdo L. The 8 Best 3D Scanning Apps for Smartphones and IPads in 2019. 2019. Available from: https://www.aniwaa.com/buyers-guide/3d-scanners/best-3d-scanning-apps-smartphones/ [Accessed: May 12, 2022]

[83] Wang D. The time dimension for scene analysis. IEEE Transactions on Neural Networks. 2005;**16**(6):1401-1426

[84] Li Z, Yan H, Ai T, Chen J. Automated building generalization based on urban morphology and gestalt theory. International Journal of Geographical Information Science. 2004;**18**(5):513-534. DOI: 10.1080/13658810410001702021

[85] Reimer LM, Weigel S, Ehrenstorfer F, Adikari M, Birkle W, Jonas S. Mobile motion tracking for disease prevention and rehabilitation using apple ARKit. In: Hayn D, Schreier G, Baumgartner M, editors. Studies in Health Technology and Informatics. Amsterdam, The Netherlands: IOS Press; 2021. DOI: 10.3233/SHTI210092

[86] Zhou X, Leonardos S, Hu X, Daniilidis K. 3D shape estimation from 2D landmarks: A convex relaxation approach. In: Proceedings of the 2015 IEEE Conference on Computer Vision and Pattern Recognition (CVPR), Boston, MA, USA, 7–15 June 2015. Boston, MA, USA: IEEE; 2015. pp. 4447-4455. DOI: 10.1109/CVPR.2015.7299074

[87] Javier Herrera P, Pajares G, Guijarro M, Ruz JJ, de la Cruz JM. Combining support vector machines and simulated annealing for stereovision matching with fisheye lenses in forest environments. Expert Systems with Applications. 2011;**38**(7):8622-8631. DOI: 10.1016/j.eswa.2011.01.066

[88] Liu Y, Wang W, Xintao X, Guo X, Gong G, Huaxiang L. Lightweight real-time stereo matching algorithm for AI chips. Computer Communications. 2022;**199**:210-217. DOI: 10.1016/j.comcom.2022.06.018

[89] Yuan W, Meng C, Tong X, Li Z. Efficient local stereo matching algorithm based on fast gradient domain guided image filtering. Signal Processing: Image Communication. 2021;**95**:116280. DOI: 10.1016/j.image.2021.116280

[90] Wilm J, Aanæs H, Larsen R, Paulsen RR. Real Time Structured Light and Applications. Kgs. Lyngby: Technical University of Denmark (DTU), 2016 (DTU Compute PHD-2015; No. 400);

[91] Scharstein D, Szeliski R. A taxonomy and evaluation of dense two-frame stereo correspondence algorithms. International Journal of Computer Vision. 2002;**47**(1):7-42

[92] Hirschmuller H, Scharstein D. Evaluation of stereo matching costs on images with radiometric differences. IEEE Transactions on Pattern Analysis and Machine Intelligence. 2008;**31**(9):1582-1599

[93] Pajares G, de la Cruz JM, López-Orozco JA. Relaxation labeling in stereo image matching. Pattern Recognition. 2000;**33**(1):53-68. DOI: 10.1016/S0031-3203(99)00036-9

[94] Ma X-L, Yuan R-Y, Zhang L-B, He M-Y, Zhang H-L, Xing Y, et al. Augmented reality autostereoscopic 3D display based on sparse reflection array. Optics Communications. 2022;**510**:127913. DOI: 10.1016/j.optcom.2022.127913

[95] Pajares G, de la Cruz JM. On combining support vector machines and simulated annealing in stereovision

matching. IEEE Transactions on Systems, Man, and Cybernetics. Part B, Cybernetics. 2004;**34**(4):1646-1657. DOI: 10.1109/tsmcb.2004.827391

[96] Dive into the world of augmented reality [Internet]. 2022. Available from: https://developer.apple.com/augmented-reality [Accessed: October 14, 2021]

[97] RealityKit. Simulate and render 3D content for use in your augmented reality apps [Internet]. 2022. Available from https://developer.apple.com/documentation/realitykit [Accessed: October 15, 2021]

[98] Rosenfeld A, Hummel RA, Zucker SW. Scene labeling by relaxation operations. In: IEEE Transactions on Systems, Man, and Cybernetics. Vol. SMC-6. June 1976. no. 6. pp. 420-433, DOI: 10.1109/TSMC.1976.4309519

Chapter 6

PID-like Fuzzy Controller Design for Anti-Slip System in Quarter-Car Robot

José R. García-Martínez, Edson E. Cruz-Miguel, Juvenal Rodríguez-Reséndiz, Luis D. Ramírez-González and Miguel A. Rojas-Hernández

Abstract

The design strategy of an adaptive Proportional-Integral-Derivative (PID)-like fuzzy controller for an anti-slip Quarter-Car robotic system is proposed. The proposed control system is constructed by two loops, an external one for lineal speed control and an internal loop for current control. A motion profile is used to follow a trajectory. The slip is computed, such as the difference between the linear velocity given by an S-curve velocity profile and the longitudinal speed calculated according to the rotational speed of the Quarter-Car tire. This difference is the input of the external control loop. Whether the slip is significant, the slave controller must do that both velocities go at the same speed controlling the current of the direct current (DC) motor. On the other hand, the mathematical model of a tire coupled to the DC-motor model is presented to simulate the system and controller response. To test the robustness of the system, different scenarios are presented where the slip coefficient varies depending on the work surface. Three surfaces were selected to test the performance of the controller, dry, wet, and icy surfaces, while the system had a trajectory.

Keywords: fuzzy control, slip-control, traction controller, trajectory planning, quarter-car robot

1. Introduction

The human being shows the difficulty in making decisions when there is imprecise information. Fuzzy logic, developed by Lofti Asker Zadeh in 1965, allows emulating of human reasoning and making correct decisions despite the information [1, 2]. It is considered a flexible tool that is based on linguistic rules dictated by experts, composed of a set of mathematical principles based on degrees of membership, whose function is to model information [3]. This modeling is done based on linguistic rules that approximate a function through the relationship between the inputs and outputs of the system. This logic presents membership ranges within an interval between zero and one, unlike conventional logic, in which the content is limited to two values: zero and one [4].

Control systems, on the other hand, are an arrangement of physical components linked or related in such a way that they command, direct or regulate the same system or another [5]; control systems are classified as open-loop control systems and [6] closed-loop control systems. Systems in which the output has no effect on the control action are called open-loop control systems. Systems that maintain a given relationship between the output and the reference input, comparing them and using the difference as a means of control, are called closed-loop control systems or feedback control systems [7, 8].

Fuzzy control is created from the combination of fuzzy logic and control systems techniques, which can be considered an expert closed-loop system in real time, implemented from the experience of an operator or process engineer unfamiliar with the process. It lends itself to being easily expressed in situation-action rules instead of differential equations [9].

Controllers based on fuzzy logic or fuzzy systems are represented by propositional rules *if-then* that can provide an understandable and easy-to-use knowledge representation [10]. This can be seen as a high-level programming language, where the program consists of conditional rules and the compiler or interpreter results in a nonlinear control algorithm, so programming through qualitative statements, represented by of *if-then* statements, to obtain a program that works in quantitative domains, provided by signals from sensors and actuators is the basis of fuzzy control [11]. Intuitively, this implies a loss of information, because there is no single translation from a qualitative entity to a quantitative representation, except in some special cases.

Traction control systems (TCSs) have come to revolutionize the behavior of automobiles as we know them today; however, thanks to the performance shown in human-crewed vehicles, it is possible to open a branch of the study of the traction controllers for mobile robots. The implementation of traction controllers in mobile robots seeks to improve the performance of the robots either due to energy consumption or the accuracy of arrival at the desired point [12]. The energy consumption in mobile robots depends significantly on the adhesion between the tires and the ground; therefore, the greater the bonding, the better stability in autonomous driving produced. This can be translated into more excellent stability in the system. Traction, on the other hand, is considered a vehicular propulsive force produced by the friction present between the tire and the surface. The inherent friction characteristics are nonlinear and uncertain, making TCSs have a high degree of [13] design.

In robotics, these types of systems are not relatively new but have not been fully addressed due to the scarcity of processing tools; current advances in embedded systems have allowed techniques from different areas to be implemented in [14] robotics. TCSs applied to mobile robots are based on modern controllers and nonlinear sliding mode controllers based on state observers [13–15].

Being wholly linked to friction, traction exhibits nonlinear behavior over time, so modern controllers must append a matrix representing the approximate nonlinear coefficients of friction to act on the control signal. In combination with traction controllers, mobile robots allow the robot to move freely on spaces with smooth, wet, and slippery surfaces. Traction control lets the robot wheels turn at similar speeds on surfaces with near-zero coefficients of friction, and when an imbalance exists, the traction controller must be activated to prevent unnecessary plant slippage, which translates to not reaching the working path [16].

Some authors have worked on different strategies related to slip control and traction control for mobile robots. For instance, [17] introduces an adaptive control strategy for a tracked mobile robot that compensates for the longitudinal slip to reach a trajectory. Another controller based on the dynamic and kinematic model with slip is presented in [18]. [19] proposes an algorithm for optimal slip control of wheeled robots with the trade-off between traction and energy consumption based on observing a change in a robot's velocity on different soil surfaces. An optimal slip ratio control using a current sensing method is presented in [20]. The controller consisted of a fuzzy PID structure. In [21] shows a slip control for a nonholonomic wheeled robot where the kinematic model of a differential-driven wheeled mobile robot is used to solve the trajectory tracking problem using fuzzy and optimal fuzzy logic. The trajectory tracking problem of a wheeled mobile robot which is actuated by two independent electrical motors is attacked in [22]. A simple non-mode-based fuzzy logic controller is used to reduce the tracking error provoked by the slippage.

As the reader can see, there are many works with different perspectives to attack the slip control in other structures. Hence, this work proposes a new methodology to control the slippery of a Quarter-Car robot using an internal loop based on fuzzy logic inference to compute the gains of a Proportional-Integral (PI) structure. The slip is calculated, such as the difference between the linear velocity given by an S-curve velocity profile and the longitudinal speed calculated according to the rotational speed of the Quarter-Car tire. This difference is the input of the external control loop. Whether the slip is significant, the slave controller must do that both velocities go at the same speed controlling the current of the DC motor. The external loop controls the angular velocity so that the linear velocity has the same magnitude. The outer loop uses an adaptive PID-like fuzzy controller structure.

The chapter is divided into the following sections. Section 2 describes the dynamic model of the test bench. Section 3 presents the methodology for self-tuning PID-like fuzzy controller. In Section 4, the simulations and results of the experimentation are presented, and, finally, in Section 5, the conclusions are given.

2. Dynamic model of one-wheel robot

This section presents the fundamental parts that compound the Quarter-Car robot structure. **Figure 1** shows the platform used for the project simulations. The platform consists of a wheel–motor coupling that will emulate the behavior of a mobile robot

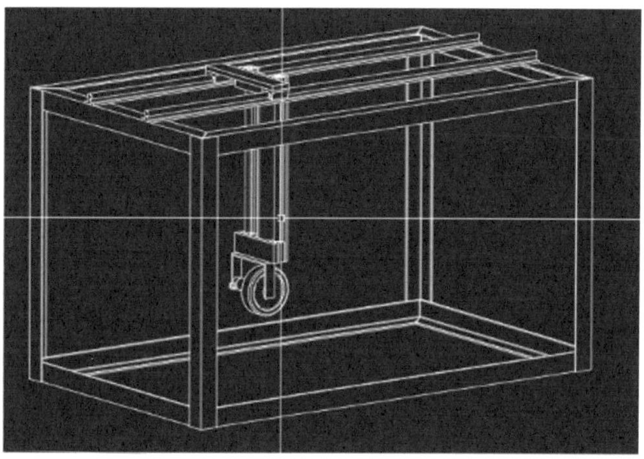

Figure 1.
One-wheel robot for control tests.

axis when different surfaces interact. The measurements will be carried out: angular velocity, longitudinal velocity, current, and slip. Slippage will be controlled so that the tire does not slip and can correctly follow a trajectory.

2.1 DC motor

Direct current (DC) motors are the most common actuators within control systems. It directly provides a rotational movement, and together with the wheels, rails, and cables, it can perform a translational action. The equivalent electrical circuit of the armature and the free body diagram of the rotor are shown in **Figure 2**. It is necessary to obtain a dynamic model that allows a correct analysis. The dynamic model of this servo system depends on the electrical and mechanical characteristics, such as the resistance R_a, the inductance L_a, the inertia J of the armature, the back electromotive force v_b, and the friction D.

From **Figure 2**, it is possible to derive the following equations based on Newton's second law for rotational motion and Kirchhoff's second law, Eqs. (1) and (2), respectively.

$$\frac{d\omega_m}{dt} = \frac{K}{J} i_a - \frac{D}{J} \omega_m \tag{1}$$

$$\frac{di_a}{dt} = \frac{e_a}{L_a} - \frac{K}{L_a} \omega_m - \frac{R_a}{L} i_a \tag{2}$$

$$\frac{d\theta_m}{dt} = \omega_m \tag{3}$$

2.2 Tire

Tire friction models are also indispensable for accurately reproducing friction forces for simulation purposes. A common assumption in most tire friction models is that the normalized tire friction, Eq. (4), is a nonlinear function of the normalized

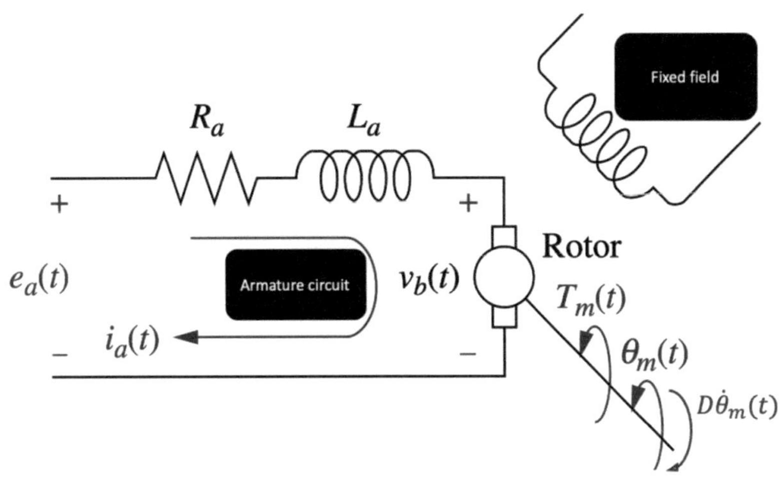

Figure 2.
Dynamic model of a DC motor.

relative velocity between the surface and the tire (slip coefficient s) with a different maximum.

$$\mu = \frac{F}{F_n} = \frac{\text{Friction force}}{\text{Normal forcel}} \tag{4}$$

Furthermore, it is understood that μ also depends on vehicle speed and road surface conditions, among other factors. This work considers the simplified motion dynamics of a Quarter-Car model. The system is represented by Eqs. (5) and (6).

$$m\dot{v} = F \tag{5}$$

$$J\dot{\omega} = -rF + u \tag{6}$$

Where m is $\frac{1}{4}$ of the mass of the vehicle, and J and r are the inertia and radius of the wheel, respectively. v is the linear speed of the tire, and ω is the angular speed of the wheel, u is the acceleration or braking torque, and F is the friction force as shown in **Figure 3**.

The most common tire friction models used in the literature are those of algebraic slip/force relationships. They are defined as one-to-one maps (memory-less) between the friction F and the longitudinal slip rate s, which is defined in Eq. (7).

$$s = \begin{cases} s_f = \dfrac{r\omega - v}{v} & \text{if } v > r\omega, \;\; v \neq 0 \;\; \text{for breaking} \\ s_m = \dfrac{r\omega - v}{r\omega} & \text{if } v < r\omega, \;\; \omega \neq 0 \;\; \text{for movement} \end{cases} \tag{7}$$

Slippage results from the reduction in the effective circumference of the rim as a result of surface deformation due to the tire rubber's elasticity. This, in turn, implies that the longitudinal velocity v will not be equal to $r\omega$. The absolute value of the slippage is defined in the interval $[0, 1]$. When $s = 0$, there is no slip (pure rotation), while $|s| = 1$ indicates total slip/skid.

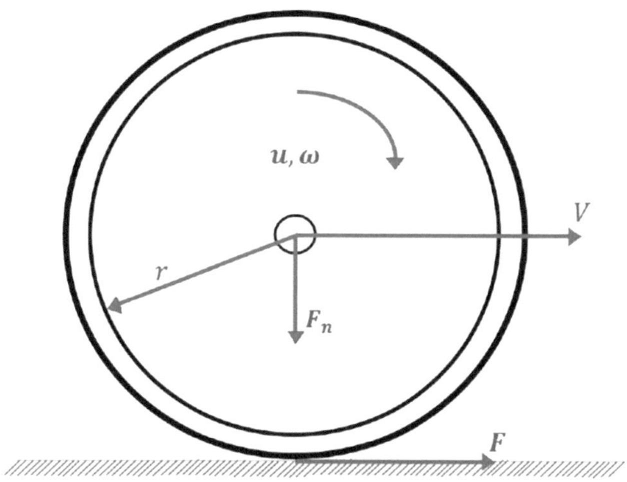

Figure 3.
One-wheel system with concentrated friction.

2.3 Slip-force models

Slip/force models aim to describe the slip movement through its force/surface dependency mapping $F(s) : s \mapsto F$. They can also depend on the vehicle speed v, that is, $F(s, v)$, and vary when the characteristics of the road change.

One of the best-known models of this type is the Pacejka model [23], also known as the "Magic Formula". It has been shown that this model agrees adequately with the experimental data obtained under particular conditions of constant linear and angular velocity. Pacejka's model is represented by Eq. (8).

$$F(s) = D\sin(\text{Carctan}(Bs - E(Bs - \arctan(Bs)))) \tag{8}$$

It can be seen that the Eq. (8) contains a set of parameters: B, C, D, and E. These parameters depend on the tire's physical properties and the vehicle's dynamic state. In the Eq. (8), D represents the maximum coefficient, C represents the shape coefficient and influences the shape of the curve, B is the stiffness coefficient, and E is the coefficient of curvature [24].

2.4 Quarter-car robot model

For the design of the traction controller, the Eqs. (1)–(3)represents the engine dynamics, and the Eqs. (5) and (6) represent the dynamics of the wheel of a differential robot.

Figure 4 shows the interaction of the dynamic equations of the motor with the wheel of the robot. It is observed that the relationship between the equations is the angular velocity of the motor shaft and the longitudinal movement of the robot, as expressed by the Eq. from (9) to the Eq. (13).

$$\frac{di_a}{dt} = \frac{e_a}{L_a} - \frac{K}{L_a}\omega_m - \frac{R_a}{L}i_a \tag{9}$$

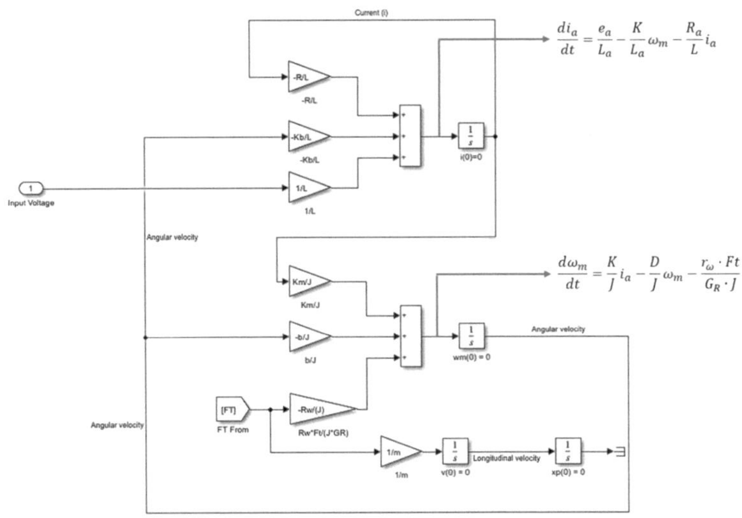

Figure 4.
Block diagram of the motor-wheel system from Eqs. (9)–(13).

$$\frac{d\omega_m}{dt} = \frac{K}{J} i_a - \frac{D}{J}\omega_m - \frac{r_w \cdot Ft}{G_R \cdot J} \tag{10}$$

$$\frac{d\theta_m}{dt} = \omega_m \tag{11}$$

$$\frac{dv}{dt} = \frac{Ft}{m} \tag{12}$$

$$\frac{dx_p}{dt} = v \tag{13}$$

The slip ratio is presented in Eq. (14) and is a function of the angular velocity.

$$s = \frac{\omega_b - \omega_f}{|\omega_f| + tol}, \quad tol \simeq 0 \tag{14}$$

On the other hand, rewriting the Eq. (14) as a function of the radius of the wheel (r_w) and the longitudinal velocity (v).

$$s = \frac{\omega r_w - v}{|\omega r_w| + tol}, \quad tol \simeq 0 \tag{15}$$

Where ωr_w and v represent the longitudinal speed of the wheel computed according to the radial speed times the radius of the tire and the longitudinal velocity measured, respectively. To calculate the traction force, the Eq. (16) of Pacejka and Sharp [23] is used.

$$F(s) = D\sin(\text{Carctan}(Bs - E(Bs - \arctan(Bs)))) \tag{16}$$

where B, C, D, and E are constants, s represents the slip value. **Figure 5** shows the implementation of Eqs. (15) and (16) in block diagram form.

On the other hand, if an imbalance is detected between the wheels of the robot, it is necessary to adjust the speed of the wheel that rotates at a higher rate with the one that turns at a lower speed [25]. To carry out this criterion, it is necessary to propose

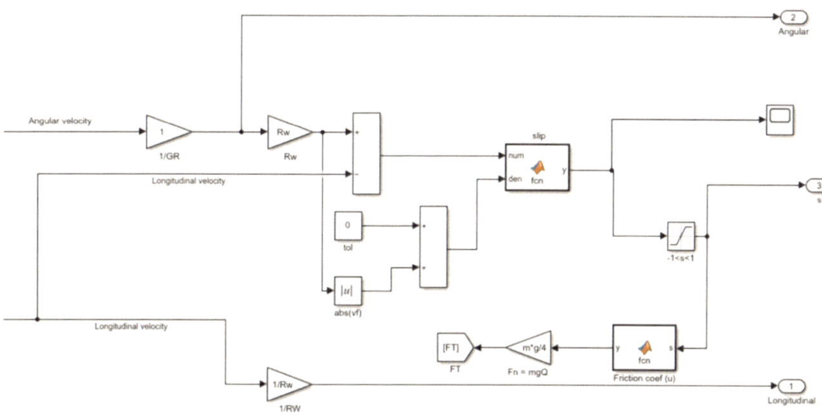

Figure 5.
Block diagram of the motor-wheel system from Eqs. (9)–(13) for the calculation of the traction force from the coefficient of friction.

behavioral variables that mathematically model the speed adjustment. In order to carry out the previous step, it is required to define a database that protects all the behavior information. This information is the linguistic variable. In addition, the operating range of the controller must be considered, which is translated into modeling the behavior in the decompensation of the wheels.

3. PID-like fuzzy controller

The design of controllers based on fuzzy logic plays an essential role in intelligent systems due to the ease of design and implementation, which must be subject to direct collaboration with the person in charge of monitoring the process to be controlled; that is, the design of the controller must be based on the experience of the system operator under certain conditions to establish the linguistic variables that the controller must obey [9, 26]. The fuzzy logic is used to tune the gains of a PID structure controller. The general form of the PID controller is depicted in Eq.(17).

$$u(t) = k_p e(t) + k_i \int_0^t e(\tau) d\tau + k_d \frac{de(t)}{dt} \tag{17}$$

This system is known as a PID-like fuzzy controller [27, 28]. The fuzzification range for computing k_p, k_d, and k_i gains, in its crisp value form, is selected according to the error, rate of change in error, and the integral of error, respectively. **Figure 6** depicts the structure proposed for the outer loop or master loop. The master loop controls the longitudinal velocity according to the angular velocity.

Figure 7 presents the structure of the internal loop. It consists of a PI structure capable of controlling the current of the tire coupled to the DC motor. Whether there is an imbalance in the speeds, it means that there is a skid in the wheel. The current must be reduced to equalize the angular velocity with the longitudinal one. That is the aim of the internal structure.

Figure 8 presents a generalized form to propose the range of operation. As one can see, the three linguistic variables contain seven linguistic values, namely negative big (NB), negative medium (NM), negative small (NS), zero (ZE), positive small (PS), positive medium (PM), and positive big (PB). A "d" and "i" are added at the beginning of each linguistic value to identify if they refer to the derivative or integral linguistic variable. The linguistic values present triangular and Gaussian shapes. For the

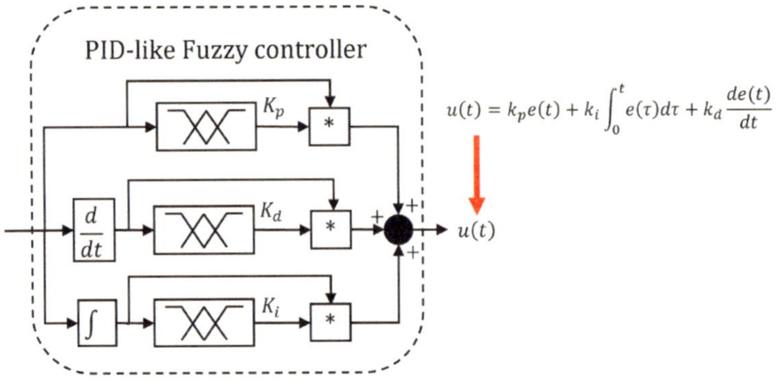

Figure 6.
Fuzzy tuner structure.

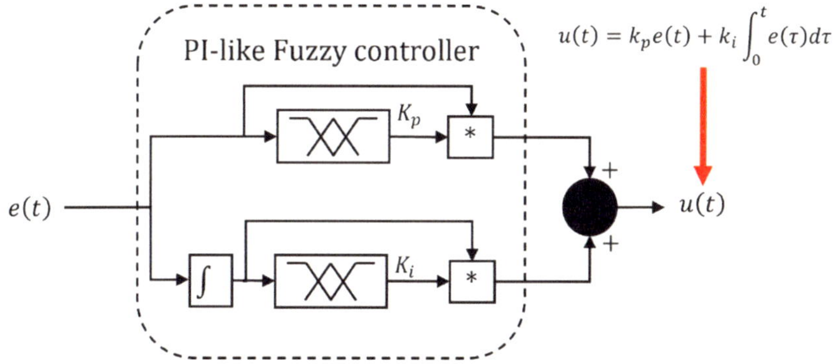

Figure 7.
Fuzzy tuner structure.

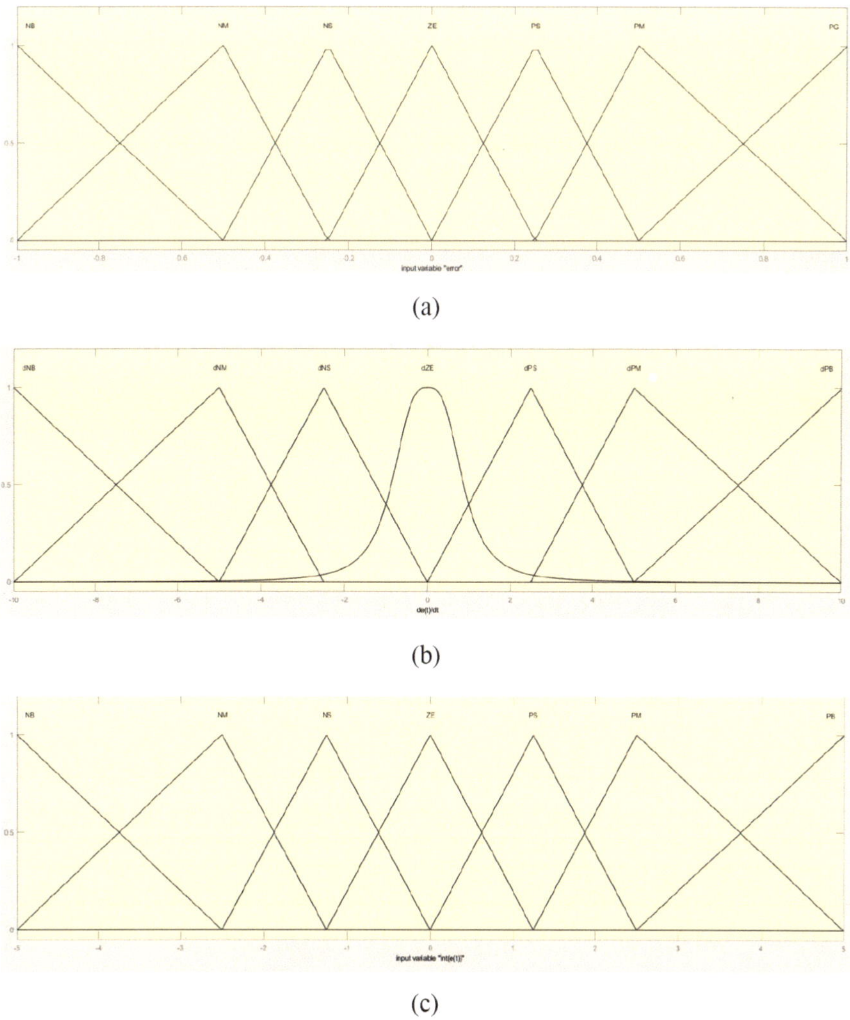

Figure 8.
Linguistic variables of the inputs (a) error, (b) derived of the error, and (c) integral of the error.

linguistic variable of error and integral of error, **Figure 8a** and **b**, respectively, triangular membership functions are used in the design since, according to the control theory, the proportional and integral actions are not as susceptible to noise. On the other hand, in the derivative error linguistic variable, **Figure 8b**, it is used a Gaussian membership function as a zero value to smooth the values of the derivative gain. The ranges of the fuzzyfication stages are $[-1, 1]$, $[-10, 10]$, and $[-5, 5]$, for k_p, k_d, and k_i, respectively.

Tables 1–3 present the fuzzy associative matrices (FAM) for computing k_p, k_d, and k_i. The inference process corresponds to a one-to-one fuzzy relationship. Likewise, it can be seen that four-linguistic values NM, NS, PS, and PM map to a single value S to calculate the output value for k_p. This relationship is used to ensure that the gain values work within a suitable range of values. The value of K_p never takes the zero valuc in this controller, only an almost zero (AZ) linguistic value. Generally, when the present error is obtained, it enters the fuzzification stage and is evaluated for the rules shown in **Table 1**. The range of values the proportional gain can take is shown in **Figure 9a**. k_p can bring any value between [1.9,9.8] and depends on the error at an instant of time. It is important to mention that k_p must not be negative for any reason.

Similar to k_p, k_d maps four linguistic values, namely dNM, dNS, dPS, and dPM map to a single value S. k_d can be zero (Z) according to the derivative error. The idea of giving a zero value to k_d is to reduce the oscillation caused by the slight variation of the change in error. The range proposed for the k_d gain goes from [0,1.4].

e(t)/kp	
NB	B
NM	S
NS	S
ZE	AZ
PS	S
PM	S
PB	B

Table 1.
FAM of k_P.

$\frac{de(t)}{dt}/k_d$	
dNB	**B**
dNM	S
dNS	S
dZE	Z
dPS	S
dPM	S
dPB	B

Table 2.
FAM of k_d.

$\int e(t)dt/k_i$	
iNB	**B**
iNM	M
iNS	S
iZE	Z
iPS	S
iPM	M
iPB	B

Table 3.
FAM of k_i.

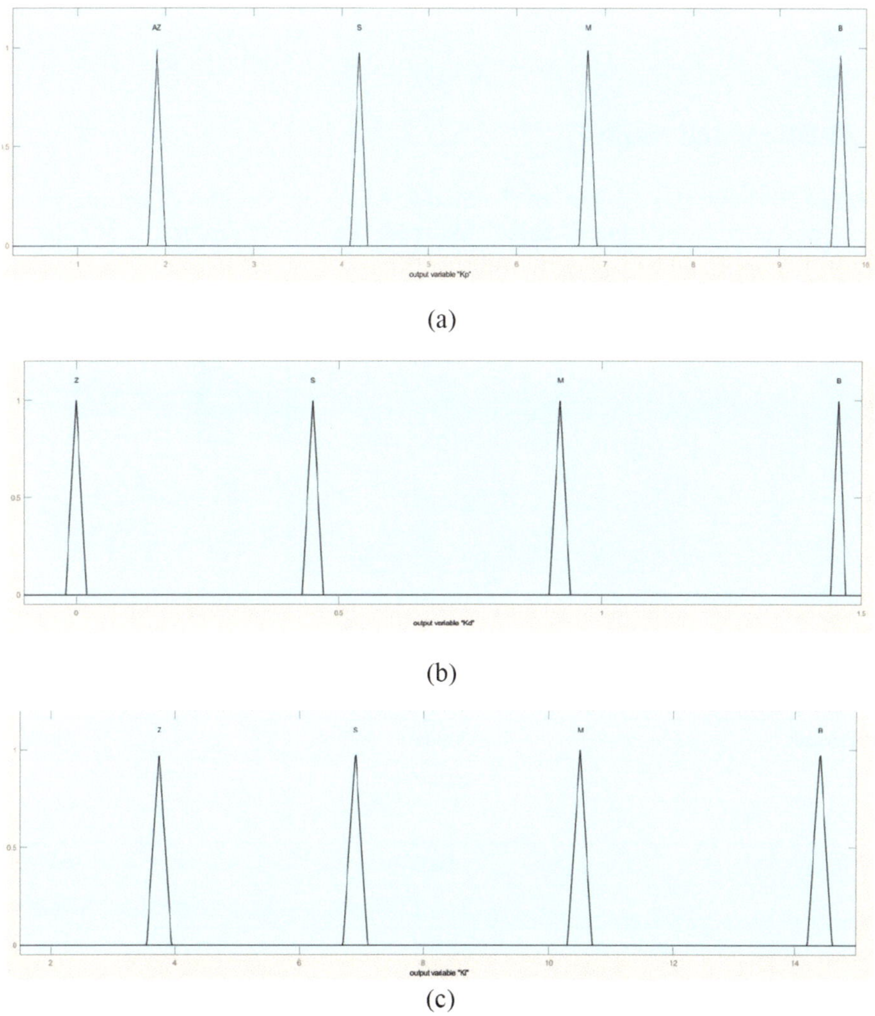

Figure 9.
Linguistic variables for the output (a) k_p, (b) k_d, and (c) k_i.

k_i does not present any alteration in how it computes its value. The range of existence is [3.5,14.5]. The integral action eliminates the steady error; for this particular case, if k_i is big enough, it compensates the error. If the error is small, the integral of the error will also reduce its magnitude considerably.

The singleton is used as a membership function in the defuzzification phase to reduce the computational cost when searching for the profit values. It is important to respect this consideration since there is only one defuzzification phase in the conventional fuzzy controller. In contrast, the proposal for this controller consists of three fuzzy steps, one for each control gain. It is necessary to mention that this process is the same for computing k_d and k_i. The centroid method, presented in Eq. 18, is used in the defuzzification stage of each gain.

$$K_{p,d,i}[n] = \frac{\sum_{i=1}^{n} \mu_c(z_i) \cdot z_i}{\sum_{i=1}^{n} \mu_c(z_i)} \tag{18}$$

where $\mu_c(z_i)$ represents degree of membership function, and z_i the position of the singleton.

4. Simulations and results

Figure 10 shows the diagram of the model, which outputs the wheel axis's slip, current, and longitudinal speeds calculated from the wheel's radius and angular velocity. **Tables 4** and **5** display the values of the parameters used for the simulation.

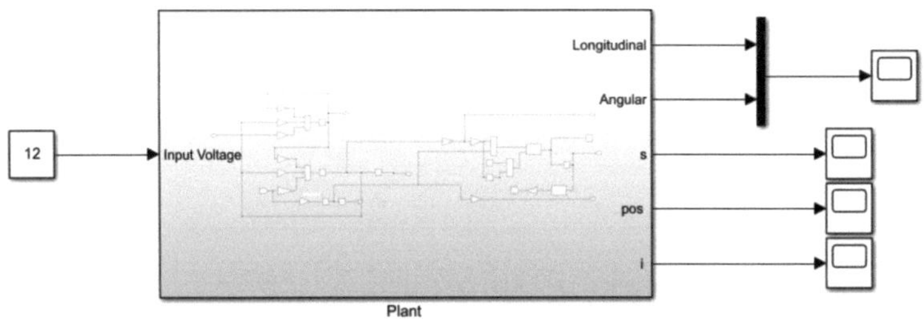

Figure 10.
Diagram model for simulation.

Parameter	Value	Units
Ra	$1.41e^{-1}$	Ω
L_a	$3e^{-3}$	H
K_τ	$5.74e^{-3}$	Volt/rad
K_e	$5.74e^{-3}$	$N \cdot m/Amp$
b	$3.97e^{-6}$	$N \cdot m \cdot s$
J	$1e^{-4}$	$kg \cdot m$

Table 4.
Parameters of the DC motor.

Parameter	Value	Units
G_R	2.5	—
Q	0.37	—
r_w	$3.2e^{-2}$	m
tol	$1e^{-10}$	—
m	1.14	kg

Table 5.
Parameters of the robot's structure.

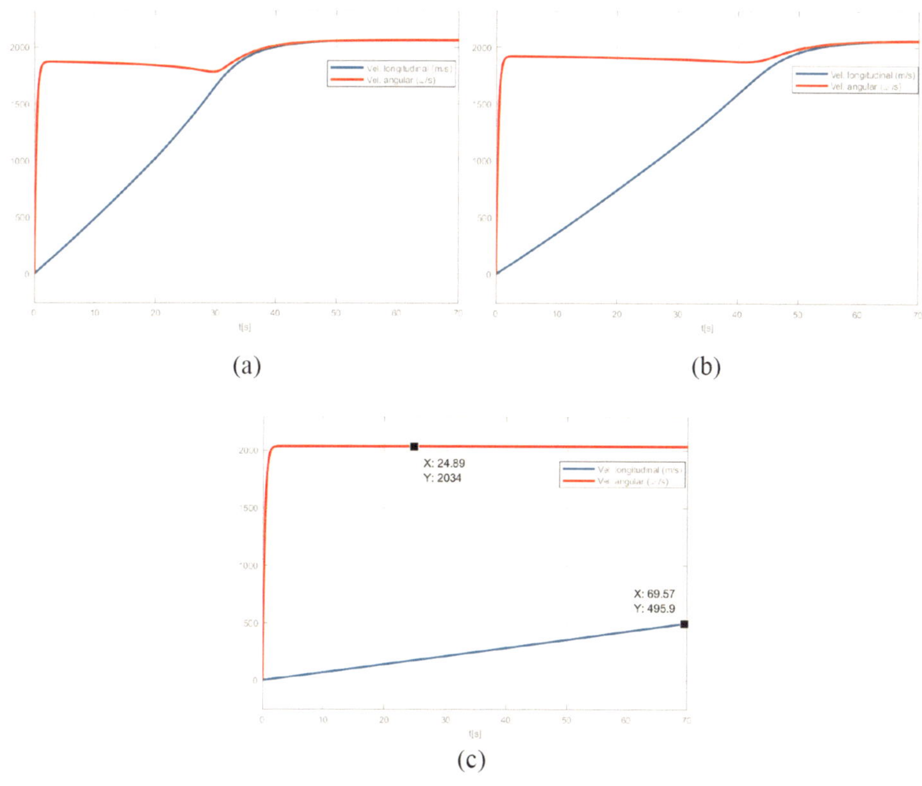

Figure 11.
Response of the Quarter-Car robot to surfaces (a) dry asphalt, (b) wet asphalt, and (c) ice.

The motor values correspond to a DC motor. **Table 5** shows the values that were used for the mechanical structure of the robot.

Assuming that the motor that is coupled to the robot's wheel is fed at 12 V and only the speeds and the slip are measured, we can say that the system is an open-loop representation, whose behavior is shown in **Figure 11** for each of the surfaces presented in **Table 6**. In **Figure 11a**, it can be seen that when the tire crosses a dry asphalt surface, the longitudinal velocity increases considerably and tries to equalize the angular velocity. On the other hand, when the tire crosses a wet asphalt surface, the longitudinal velocity takes longer to equal the angular velocity, as shown in **Figure 11b**. Finally, the angular velocity tends to its maximum permissible value when

Surface	*B*	*C*	*D*	*E*
Dry asphalt	19.25	1.65	0.92	0.6
Wet asphalt	15.8	1.6	0.62	0.6
Ice	9.8	1.45	0.1	0.6

Table 6.
Parameters for computing the coefficient of friction depending on the surface.

the tire makes contact with an ice-covered surface. In contrast, the longitudinal rate grows negligibly as a function of time; see **Figure 11c**. This is because the coefficient of friction is very close to zero, which means that the friction is zero, causing the tire to tend to skid without moving considerably longitudinally.

The Algorithm 1 presents the way to compute the friction coefficient used for simulation.

Algorithm 1:

Calculation of the coefficient of friction.
Función y = fcn(s)
B;
C;
D;
E;
if $s < 0$ **then**
  s = abs(s)
  $\mu = -D\sin(C\arctan(Bs - E(Bs - \arctan(Bs))))$
else
  $\mu = D\sin(C\arctan(Bs - E(Bs - \arctan(Bs))))$
end
y = μ;

The behavior of the robot as a function of longitudinal and radial speed is presented in **Figure 12**, which contains the behavior over time of the difference in rates, better known as sliding. As shown in **Figure 11**, the speeds of the wheel have different magnitudes. This means that the angular speed is much more significant in the first moments than the longitudinal speed. This behavior is without any controller, so there is no device that regulates the speeds from the beginning of the operation. However, in **Figure 12**, it can be seen that the slip tends to zero when the speeds reach the same magnitude on dry and wet asphalt surfaces, **Figure 11**, while for the icy surface, there is always a slip value, and the tire remains rotating without moving longitudinally.

4.1 PID-like fuzzy controller results

For the design of the PID-like fuzzy controller, the methodology presented in the section 3 was used; see **Figure 13**.

In order to solve the problem of different speeds, a slip controller will be designed whose reference input is from a seven segments motion profile. The wheel–motor

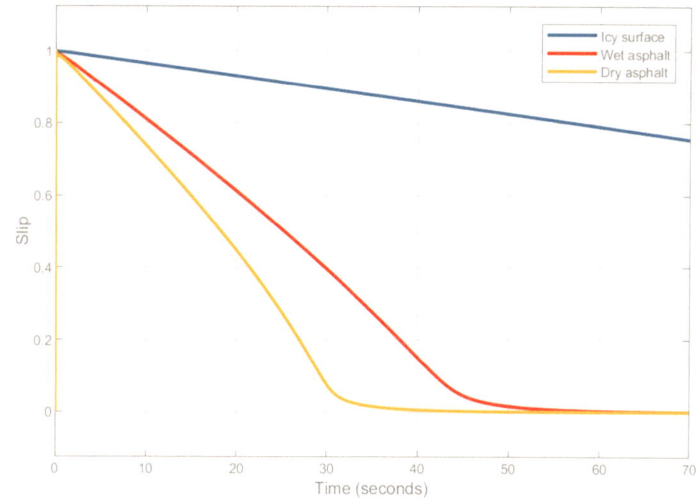

Figure 12.
Slip response to a 12 V input.

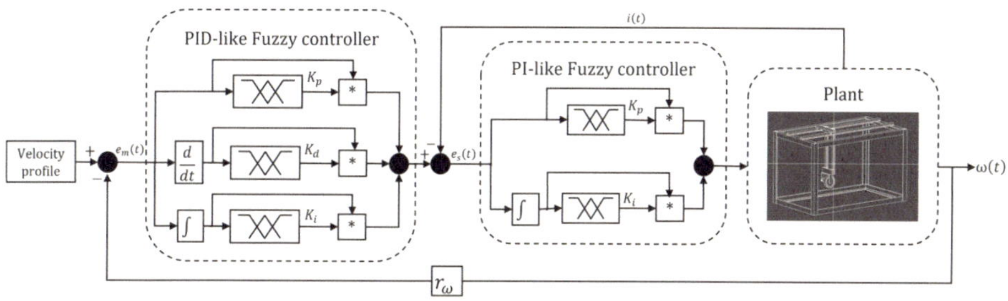

Figure 13.
PID-like fuzzy controllers and the Quarter-Car robot structure.

system must ensure the same angular and longitudinal speed by following the change in reference speed at any time. In **Figure 14**, the velocity profile is shown as input. It goes through two integrators because the jerk profile is being generated, so when integrating once, you get the reference acceleration, and the second time you get the input velocity. The longitudinal velocity is compared concerning the motion profile. The idea is that the rim moves according to the reference and can work with the surfaces presented in **Table 6**. The difference between the reference speed and the longitudinal speed produces the error, which is the input to the master controller. The

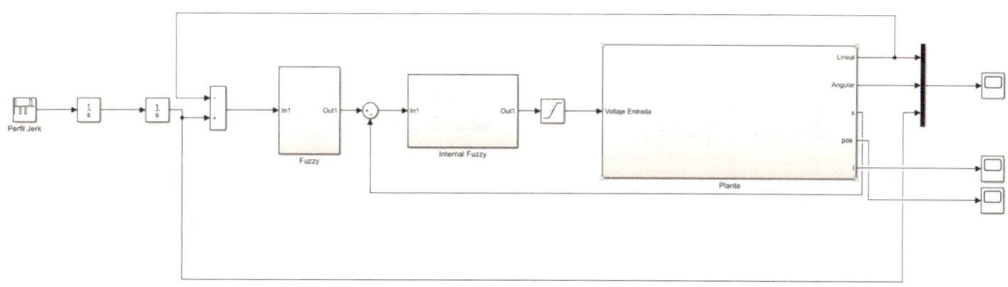

Figure 14.
Fuzzy controller.

output of this controller is compared to the slip to generate an internal control loop to prevent slippage.

The Quarter-Car and the control system used for the simulation are shown in **Figure 14**. The diagram shows a cascade controller with a longitudinal speed master loop and a current slave loop. Both control structures are based on fuzzy logic. For the master loop, a self-tuning PID controller is used. This structure is shown in detail in **Figure 15**. The same rules were used and the same number of linguistic values. The difference lies in the range of operation of the linguistic variables of gains k_p, k_d, and k_i. These ranges were proposed from the PID controller answer from the previous section. For the slave loop, a PI structure with self-tuning was proposed, and the same strategy as the one mentioned above was used. Only the range of values of the input to the internal k_p and k_i profit parts has been modified.

The ranges of values are smaller and are intended to compare the response of the master controller with the slip produced by the tire and the selected surface to evaluate.

The simulation result is shown in **Figure 15**, where each surface shows a similar behavior when following the reference and where the angular and longitudinal velocities do not offer a significant displacement. This is to the slide controller function. For example, in **Figure 15**, a dry asphalt surface is used, and it can be seen that the speeds adhere to the reference. In 15–*b*, it can be seen that the longitudinal velocity tends to

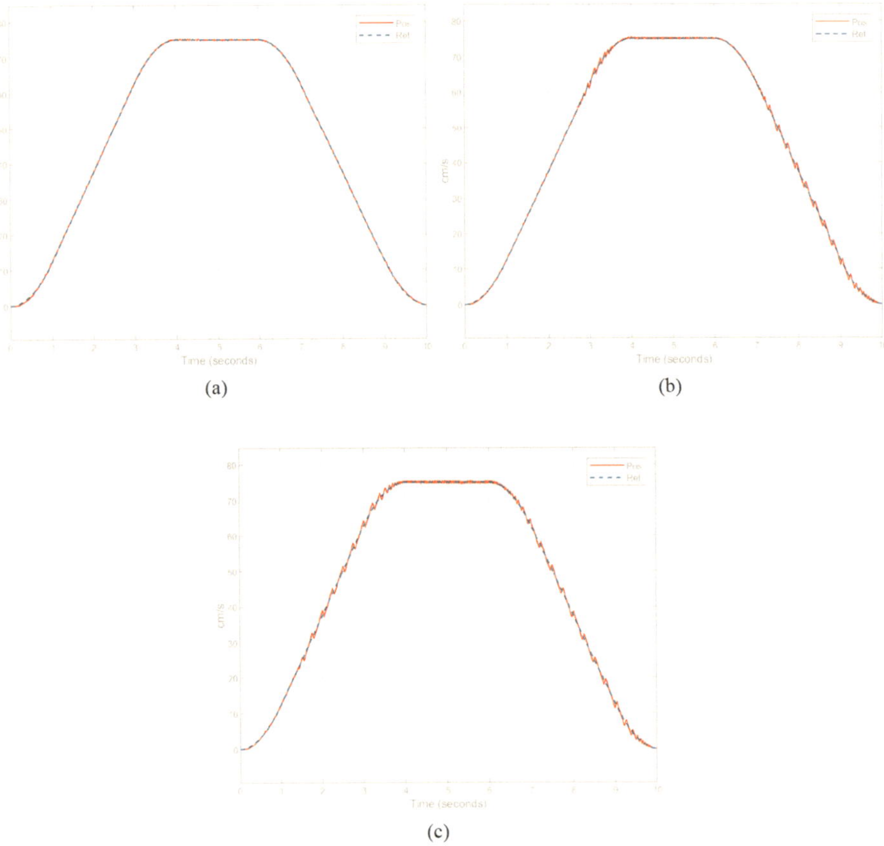

Figure 15.
Response of the tire-motor plant to surfaces (a) dry asphalt, (b) wet asphalt y, (c) ice.

decrease in magnitude, although the change in it is not significant. Finally, **Figure 15c** shows the behavior of the angular and longitudinal velocities on an ice surface. It is shown that the controller is reacting well, the longitudinal speed reduces its magnitude, and the angular speed follows the profile. The latter makes sense since the slip functions the angular and longitudinal velocities. Therefore, the magnitude of the velocities must be similar depending on the surface over which the plant moves.

The slip measured by the simulation is shown in **Figure 16**. As can be seen, there is less slip for dry asphalt. This was to be expected since this type of surface provides a better grip on the tires to the surface. On the other hand, we see that slip tends to increase in magnitude when the robot's wheel moves on a wet surface, similar to a larger vehicle when it rains. Finally, when the surface contains ice, the landslide tends to increase in magnitude since surfaces with ice have a minimal coefficient of friction, causing the landslide to increase in magnitude if it is not controlled.

The current measured in the simulation is shown in **Figure 17**. The dry asphalt surface presents a greater magnitude of current mainly due to the coefficient of friction since there is always a grip; that is, there is a considerable friction force. When the tire runs on a wet surface, it tends to decrease the coefficient of friction, making this one turn more efficiently, thus reducing the current supplied to the motor. Finally, when the surface is ice, the current tends to decrease since the wheel will tend to slip, which means that the amount of current that is going to be injected into the motor is less, and in order to compensate for the speed in the wheel or on the motor shaft.

The final position or target position is deduced from the 7-segment profile. **Figure 18** shows the position of the system with its respective reference. **Table 7** displays the values of the desired and measured positions of the different surfaces.

Table 8 shows the performance of the controllers against their respective surfaces using the root mean square error (RMSE). As can be seen, the PID controller shows a better performance than the fuzzy controller on a dry asphalt surface. The adaptive

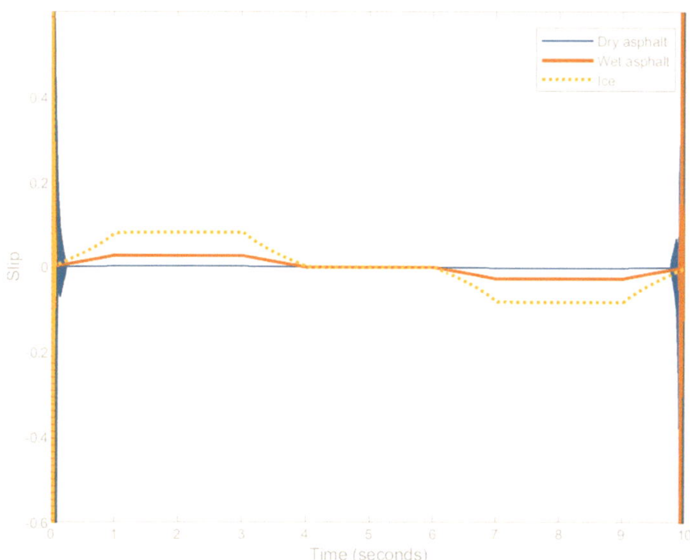

Figure 16.
Slip behaviour.

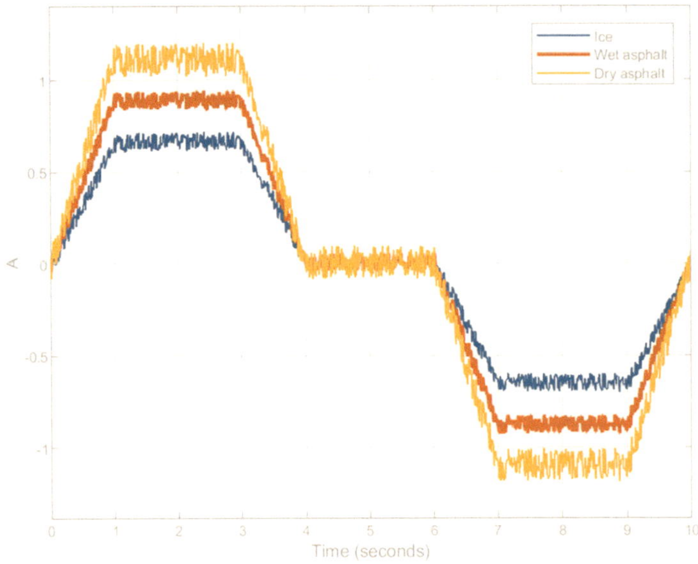

Figure 17.
Current consumption.

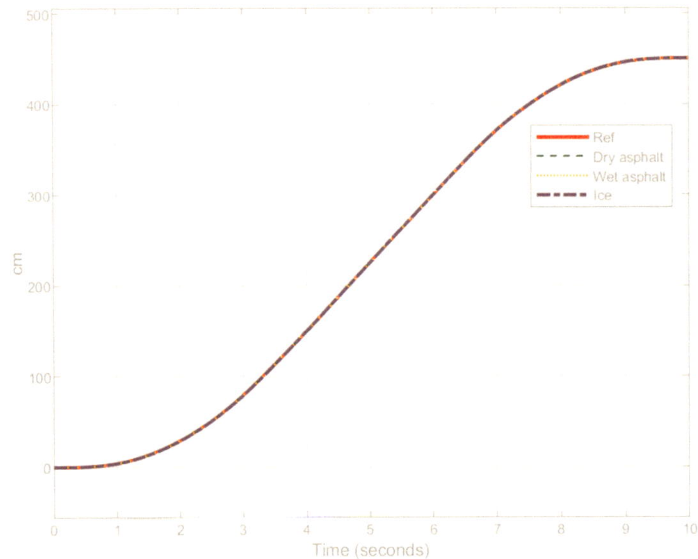

Figure 18.
Perfil de 7 segmentos.

Reference	Dry asphalt	Wet asphalt	Ice
450 cm	450 cm	451.05 cm	455.2 cm

Table 7.
Final position values.

Structure	Dry asphalt	Wet asphalt	Ice
PID	0.01498	0.01673	593.9
Fuzzy controller	0.01577	0.01658	0.01913

Table 8.
Performance values.

PID-like fuzzy controller starts to increase its performance when there is a change of surface, for example, for the wet surface. Finally, better performance is appreciated when the surface is icy since the traction controller allows longitudinal speed to be controlled by controlling the wheel's speed, making the latter follow the reference.

5. Conclusions

The mathematical model of the motor and the wheel are used for the traction controller. The angular speed, longitudinal speed, current, and slip are obtained from this model. The current is used in the slave loop, whose objective is to brake the wheel to compensate for the angular longitudinal speed through the motor shaft. The master loop aims to follow an S-curve velocity profile. Three different surfaces were used for the simulations: dry asphalt, wet asphalt, and ice.

A PID controller was implemented to make the comparisons, before the surfaces mentioned above, with the fuzzy controller. The fuzzy controller design followed the same methodology as the one used for the motion controller. This makes the adaptive PID-like fuzzy controller master–slave design methodology easy to reproduce. The PID controller worked well on wet and dry asphalt surfaces, but the controller no longer compensated for speeds when the Quarter-Car robot was presented on an icy surface. The same gains were used for all three trials. The fuzzy controller worked well for all three surfaces, showing robustness to changing surfaces. For future work, a type-2 fuzzy logic controller will be implemented to prove the type reduction algorithms in embedded systems in the slip control applications.

Conflict of interest

The authors declare no conflict of interest.

Abbreviations

PID	Proportional-integral-derivative
DC	Direct current
TCS	Traction constrol system
FAM	Fuzzy associative matrices
NB	Negative big
NM	Negative medium
NS	Negative small
ZE	Zero
PS	Positive small

PM	Positive medium
PB	Positive big
S	Small
AZ	Almost zero
B	Big
RMSE	Root mean square error

Author details

José R. García-Martínez[1*†], Edson E. Cruz-Miguel[1†], Juvenal Rodríguez-Reséndiz[2†], Luis D. Ramírez-González[1†] and Miguel A. Rojas-Hernández[1†]

1 Universidad Veracruzana, Poza Rica, Ver., Mexico

2 Universidad Autónoma de Querétaro, Querétaro, Mexico

*Address all correspondence to: romangarcia@uv.mx

† These authors contributed equally.

References

[1] Zadeh LA. A Fuzzy-Algorithmic Approach to the Definition of Complex or Imprecise Concepts. In: Systems Theory in the Social Sciences. Interdisciplinary Systems Research / Interdisziplinäre Systemforschung. Birkhäuser, Basel: Springer; 1976. pp. 202-282

[2] Alavala CA. Fuzzy Logic and Neural Networks: Basic Concepts & Application. New Delhi: New Age International; 2008

[3] Hanss M. Applied Fuzzy Arithmetic. Berlin Heidelberg: Springer; 2005

[4] Cruz PP, Herrera A. Inteligencia artificial con aplicaciones a la ingeniería. México: Alpha Editorial; 2011

[5] Distefano JJ, Stubberud AR. Retroalimentación y sistemas de control. Vol. 2. Colombia: McGraw-Hill; 1995

[6] Kuo BC. Automatic Control Systems. United States: Prentice Hall PTR; 1987

[7] Ogata K. Ingeniera de control moderna. España: Pearson Educación; 2003

[8] Nise NS. Control Systems Engineering. United States: John Wiley & Sons; 2020

[9] Mandal AK. Introduction to Control Engineering: Modeling, Analysis and Design. New Delhi: New Age International; 2006

[10] Jager R. Fuzzy Logic in Control. Amsterdam: Rene Jager; 1995

[11] Kacprzyk J. Studies in fuzziness and soft computing 295. Berlin, Heidelberg: Springer; 2013

[12] Albagul A, Martono W, Muhida R. Dynamic modelling and adaptive traction control for mobile robots. In: Kordic V, Lazinica A, Merdan M, editors. Cutting Edge Robotics. London, UK, Rijeka: IntechOpen; 2005 Chapter 1

[13] Kuntanapreeda S. Traction control of electric vehicles using sliding-mode controller with tractive force observer. International Journal of Vehicular Technology. 2014;**2014**:1-10

[14] Mu J, Yan X-G, Spurgeon SK, Mao Z. Nonlinear sliding mode control of a two-wheeled mobile robot system. International Journal of Modelling, Identification and Control. 2017;**27**(2): 75-83

[15] Konduri S, Orlando E, Torres C, Pagilla PR. Effect of wheel slip in the coordination of wheeled mobile robots. IFAC Proceedings Volumes. 2014;**47**(3): 8097-8102

[16] Alhaj Ali SM, Hall EL. Design and simulation of a motion controller for a wheeled mobile-robot autonomous navigation. In: Intelligent Robots and Computer Vision XXIII: Algorithms, Techniques, and Active Vision. Vol. 6006. International Society for Optics and Photonics; Proc. SPIE. 2005. p. 60060M

[17] Iossaqui JG, Camino JF, Zampieri DE. A nonlinear control design for tracked robots with longitudinal slip. IFAC Proceedings Volumes. 2011;**44**(1): 5932-5937

[18] Ryu J-C, Agrawal SK. Differential flatness-based robust control of mobile robots in the presence of slip. The International Journal of Robotics Research. 2011; **30**(4):463-475

[19] Kim J, Lee J. Intelligent slip-optimization control with traction-energy trade-off for wheeled robots on rough terrain. In: 2014 IEEE/RSJ International Conference on Intelligent Robots and Systems, IEEE, 2014. pp. 1938–1943

[20] Kim D-E, Yoon HN, Kim KS, Sreejith MS, Lee J-M. Using current sensing method and fuzzy pid controller for slip phenomena estimation and compensation of mobile robot. In: 2017 14th International Conference on Ubiquitous Robots and Ambient Intelligence (URAI), IEEE, 2017 pp. 397–401

[21] Falsafi MH, Alipour K, Tarvirdizadeh B. Tracking-error fuzzy-based control for nonholonomic wheeled robots. Arabian Journal for Science and Engineering. 2019;**44**:881-892

[22] Falsafi M, Alipour K, Tarvirdizadeh B. Fuzzy motion control for wheeled mobile robots in real-time. Journal of Computational & Applied Research in Mechanical Engineering (JCARME). 2019;**8**(2):133-144

[23] Pacejka HB, Sharp RS. Shear force development by pneumatic tyres in steady state conditions: a review of modelling aspects. Vehicle System Dynamics. 1991;**20**(3-4):121-175

[24] Bakker E, Nyborg L, Pacejka HB. Tyre modelling for use in vehicle dynamics studies. SAE Transactions. 1987:190-204

[25] Jin L, Chen P, Zhang R, Ling M. Longitudinal velocity estimation based on fuzzy logic for electronic stability control system. Advances in Mechanical Engineering. 2017;**9**(5): 16878140176986s62

[26] De Silva CW. Intelligent Control: Fuzzy Logic Applications. Boca Raton: CRC Press; 2018

[27] García-Martínez JR, Cruz-Miguel EE, Carrillo-Serrano RV, Mendoza-Mondragón F, Toledano-Ayala M, Rodríguez-Reséndiz J. A pid-type fuzzy logic controller-based approach for motion control applications. Sensors. 2020;**20**(18):5323

[28] Doan D-V, Nguyen K, Thai Q-V. Load-frequency control of three-area interconnected power systems with renewable energy sources using novel pso pid-like fuzzy logic controllers. Engineering, Technology & Applied Science Research. 2022;**12**(3):8597-8604

Chapter 7

Methodology for the Implementation of a Fuzzy Controller on Arduino, MATLAB™ and Nexys 4™ Platforms

Jesus de la Cruz-Alejo, Hugo Beatriz-Cuellar, Agustin Mora-Ortega and Maria Belem Arce-Vazquez

Abstract

This chapter presents a methodology to implement a fuzzy controller in different hardware platforms, which can be used to control a system or process. The methodology proposes a programming algorithm to implement a fuzzy controller on the Arduino UNO, Arduino DUE, Nexys 4™, and MATLAB™ platforms. The programming algorithm uses two control statements (IF-THEN and FOR) and the basic mathematical operations. The fuzzy controller was designed for two input variables, one output variable, five fuzzy sets for each variable, and a Mamdani type structure. An analysis of convergence time, amount of memory, and control surface is performed to ensure that the fuzzy controller on all platforms is satisfactory. MATLAB™ is used to compare these platforms through numerical simulations, which demonstrates the effectiveness of the proposed methodology. The experimental results of the fuzzy controller are a processing time of 117 milliseconds and 40% of the memory of the Arduino UNO, a processing time of 21.275 milliseconds and 5% of the memory of the Arduino DUE, and a processing time of 17.871 milliseconds and 40% of the memory on the Nexys 4™. Finally, a Mean Square Error of 0.0326, 0.0643, and 0.1125 was obtained for MATLAB™, Arduino, and Nexys 4™, respectively.

Keywords: fuzzy controller, methodology, MATLAB™, Arduino, Nexys 4™

1. Introduction

Currently, most of the processes use a control system, which provides the necessary conditions and guarantees the correct operation of the process to obtain the final product. Additionally, there is a great variety of control systems, for example, neural networks, PID controllers, robust control, sliding modes, PLC (Programmable Logic Controllers), fuzzy controller, among others. On the other hand, the characteristics of the process must be analyzed to select a control system, for example, the cost of the control system, the software or hardware used for the implementation of the control system, the mathematical requirements used to analyze the process, variables, types of sensors and actuators necessary to control the process, desired precision in the process, advantages and/or disadvantages of the control system, among other things. Therefore, this work shows a methodology for the implementation of a fuzzy controller in different software or hardware platforms, since a fuzzy controller does not need the mathematical model of the system, uses the experience or knowledge of a person, does not use complex mathematical equations for its implementation, and uses linguistic explanations (low, high, hot, cold, good, bad, etc.) to define process conditions and control action. Therefore, a fuzzy controller is one of the best options for controlling a process. Currently, fuzzy controllers are used in a wide variety of processes or applications; for example, modeling and simulation of the Maximum Power Point Tracking (MPPT) in photovoltaic solar energy systems [1–5], increase the accuracy in determining the degree of diabetes in a person [6, 7], identification of hot spots and analysis of the intensity of flames in pipes to prevent fires [8], improve the performance of a grid-connected wind generator system [9–11], control of the output voltage of a Boost converter [12], generate a suitable microclimate for an agricultural greenhouse [13], among other applications. On the other hand, the proposed methodology uses two control statements (IF-THEN and FOR), and the basic mathematical operations (addition, subtraction, multiplication, and division) for the design and implementation of the fuzzy controller stages. Therefore, the proposed methodology uses basic programming elements, which allows the fuzzy controller to be implemented in different software or hardware platforms. In this work, MATLAB™ and the Arduino UNO, Arduino DUE, and Nexys 4™ boards are used to show the correct operation of the proposed methodology. Also, Fuzzy Logic Toolbox™ is used to simulate and analyze the operation of the fuzzy controller. Finally, the structure of the chapter is as follows, Section 2 shows the procedure to implement the fuzzy controller in the different platforms, Section 3 shows the simulation and experimental results of the fuzzy controller, which was implemented in the different platforms, and Section 4 presents the conclusions.

2. Description of the methodology

The methodology proposes a programming algorithm, which allows implementing a fuzzy controller on different hardware and software platforms, which have different technical characteristics. In this case, MATLAB™ and the Arduino UNO, Arduino DUE, and Nexys 4™ boards are used to show the correct operation of the proposed methodology. The Arduino UNO board uses the ATmega328P microcontroller, which has a 32 KB Flash memory, a 2 KB SRAM memory, and a 1 KB EEPROM memory. In addition, this board has a 16 MHz clock speed, 14 digital input or output pins, 6 analog inputs with a 10-bit resolution, and 8 PWM outputs, and its programming language is based on the C/C++ language. The Arduino DUE board uses the SAM3X8E microcontroller, which has a 512 KB Flash memory and a 96 KB SRAM memory. In addition, this board has a clock speed of 84 MHz, 54 digital input or output pins, 12 analog inputs with a 12-bit resolution, 2 digital-analog converters, and 12 PWM outputs, and its language of programming is based on the C/C++ language [14, 15]. The Nexys 4™ board model XC7A100T-1CSG324C contains two external memories, a 128Mbit cellular RAM and a 128Mbit non-volatile serial Flash device. In addition, this board has a clock speed of 100 MHz, USB ports, an Ethernet port, a micro-SD port, a micro-USB port, a VGA port, accelerometer, temperature sensor, digital microphone, speaker

Figure 1.
Board of (a) Arduino UNO, (b) Arduino DUE, and (c) Nexys 4™.

amplifier, 16 user switches, 16 user LEDs, two 4-digit 7-segment displays, two three-color LEDs and its programming language is VHDL [16, 17]. Additionally, the ISE Project Navigator 14.7 software was used to program the Nexys 4™ board, and the Arduino IDE (Integrated Development Environment) was used to program the Arduino boards. **Figure 1** shows the Arduino UNO, Arduino DUE, and Nexys 4™ boards. On the other hand, the fuzzy controller implementation is done in the MATLAB™ Script, which is a program file that allows building a programming algorithm and provides tools for displaying graphics in two and three dimensions. Additionally, the MATLAB™ Script is used for the analysis, design, and simulation of the different stages of the controller. Also, the fuzzy controller is implemented in Fuzzy Logic Toolbox™, which is a MATLAB™ tool used to design, analyze, and simulate a fuzzy controller. Therefore, the Fuzzy Logic Toolbox™ is used to analyze the results obtained from the implementation of the controller on the different platforms [18, 19]. Finally, this boards can be used to control a process, and the selection of the board will depend on the characteristics of the process, for example, the number of variables, sensors or actuators, the cost of the board, the cost of the system, among other things.

2.1 Fuzzy controller design

The design of a fuzzy controller of the Mamdani type is carried out, which is made up of the fuzzification, inference, aggregation, and defuzzification stages. On the other hand, a fuzzy controller can use more than one input variable and can determine more than one output variable; however, a fuzzy controller needs at least two input variables and one output variable to function properly. If the number of input and output variables of the fuzzy controller is increased, then the complexity of the fuzzy controller implementation will increase. **Figure 2** shows the structure of a fuzzy controller of the Mamdani type. The design and implementation of a fuzzy controller begins with the definition of the controlled variable of the process, for example,

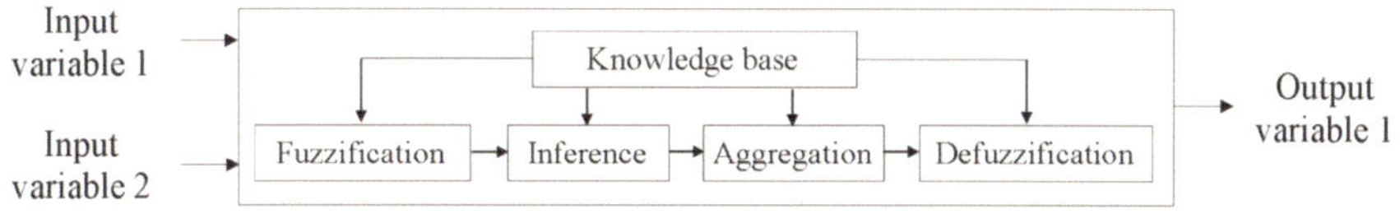

Figure 2.
Structure of a fuzzy controller of the Mamdani type.

temperature, humidity, pressure, pH, among others. Subsequently, the input and output variables of the controller must be defined. For this, it must be considered that the input variables of the fuzzy controller are used to measure the state or condition of the process, and the output variable of the fuzzy controller is the control action, which will be used to adjust the controlled variable. Also, a universe of discourse must be specified for each of the controller variable, which can be defined as the range of values, where a specific value of the input or output variables can be found or located. In other words, a universe of discourse is made up of the values that are between the minimum and maximum values of a variable. On the other hand, a series of linguistic values (low, high, good, bad, etc.) must be defined, which describe the state or condition of the input and output variables of the controller. Subsequently, within the universes of discourse of the controller variables, fuzzy sets must be defined, which must be labeled with the name of the linguistic values. Also, the type of fuzzy set that will be used to implement the controller must be defined. For this, the computational load and the necessary programming elements must be considered. Finally, these elements should be considered as the initial parameters of the fuzzy controller [20].

The fuzzy controller should be used for a specific situation since this is the correct way to show how to implement the fuzzy controller. Therefore, the fuzzy controller is used to determine the tip of a food establishment, since this application is the simplest to understand the operation of the fuzzy controller. This application will allow to characterize the controller; that is, this application allows defining the input and output variables of the controller, the length of the universes of discourse, type of fuzzy sets, dimensions of fuzzy sets, among other things. In this case, the value of the tip depends on the quality of the food and the service of the food establishment. Therefore, a signal defined as "food" and a signal defined as "service" are used as controller input variables, and a signal defined as "tip" is used as the controller output variable. A universe of discourse from 0 to 100 was used for the input variables since food and service can be evaluated with a score of 0 to 100. A universe of discussion from 0 to 120 was used for the output variable, which represents the amount of money from 0 to $120. However, the minimum value of the tip will be $20, and the maximum value of the tip will be $100.

2.2 Fuzzy sets of controller variables

There is a great variety of types of fuzzy sets, which can be used for the implementation of a fuzzy controller. **Figure 3** shows a triangular type of fuzzy set, which is used to define the fuzzy sets of the input and output variables of the controller. This type of fuzzy set uses the subtraction and division operations for its implementation in software and hardware.

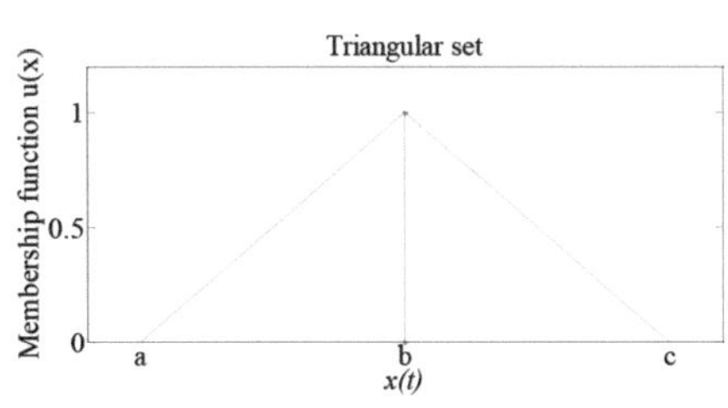

Figure 3.
Triangular fuzzy set.

$$\mu(x) = \begin{cases} 0 & \text{si } x \leq a \\ \dfrac{x-a}{b-a} & \text{si } a < x \leq b \\ \dfrac{c-x}{c-b} & \text{si } b < x \leq c \\ 0 & \text{si } x \geq c \end{cases} \tag{1}$$

On the other hand, the accuracy of the process state measurement depends on the number of fuzzy sets of the input variables. Similarly, the accuracy of the control action to adjust the process depends on the number of fuzzy sets of the output variable. Therefore, a process analysis must be performed to determine the number of fuzzy sets for the input and output variables of the controller. In this case according to the approximation error through simulations using five and seven, the approximation error between five and seven was minimal. So, five fuzzy sets were defined for the variable "food" labeled Very Bad (VB), Bad (BD), Regular (RG), Good (GD), and Very Good (VG). Also, five fuzzy sets were defined for the variable "service" labeled Very Bad (VB), Bad (BD), Regular (RG), Good (GD), and Very Good (VG). Also, five fuzzy sets were defined for the output variable "tip" labeled Very Bad (VB), Bad (BD), Regular (RG), Good (GD), and Excellent (EX). The dimensions of the fuzzy sets depend on the importance of the control action; therefore, the length of the fuzzy sets can be different. Finally, **Figure 4** shows the dimensions of the fuzzy sets of the variables of the fuzzy controller.

2.3 Fuzzification stage

The first stage of the fuzzy controller is fuzzification, which is used to transform a real variable ("food" or "service") into a fuzzy variable through the membership functions. Fuzzification determines the fuzzy sets that indicate the state or condition of an input variable. For this, the membership function μ(x) must be determined using Eq. (1) [21–23]. The algorithm used to implement the fuzzification in the

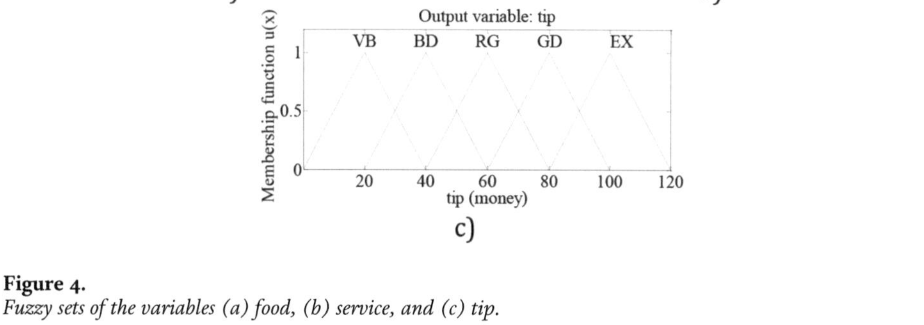

Figure 4.
Fuzzy sets of the variables (a) food, (b) service, and (c) tip.

MATLAB™ Script and on the Arduino UNO, Arduino DUE, and Nexys 4™ boards is described below.

Eq. (1) is used to determine the μ(x) of the input variables (food or service), and IF-THEN conditional statements are used to determine the fuzzy sets that define the condition of the input variables. FD and SV are the value of food and service, respectively. Setfd and Setsv store the fuzzy set that indicates the state of the input variables, Mf1 and Mf2 are the membership functions of the input variables, and a1, a2, b1, and b2 are dimensions of a fuzzy set.

% MATLAB Script	*// Arduino UNO/DUE*	*– Nexys 4™*
if FD >= a1 && FD <= b1	if (FD >= a1 && FD <= b1) {	if (FD >= a1 and FD <= b1) then
Setfd = VB;	Setfd =VB;	Setfd := VB;
Mf1=(FD – a1) / (b1 – a1);	Mf1=(FD – a1) / (b1 – a1);	Mf1:=((FD-a1)*(100))/(b1-a1);
end	}	end if;
if SV >= a2 && SV <= b2	if (SV >= a2 && SV <= b2) {	if (SV>=a2 and SV<=b2) then
Setsv = VB;	Setsv =VB;	Setsv := VB;
Mf2=(SV – a2) / (b2 – a2);	Mf2=(SV – a2) / (b2 – a2);	Mf2:=((SV-a2)*(100))/(b2-a2);
end	}	end if;

2.4 Inference stage

The inference stage uses fuzzy rules, which represent the knowledge base of a fuzzy controller and determine the controllability of the process. Fuzzy rules relate the membership functions of the fuzzy sets that the input variables have. The result of a fuzzy rule is a fuzzy set contained in the output variable obtained using the Mamdani implication. The membership functions of the input variable fuzzy sets indicate the state or condition of the process, and the fuzzy set of the output variable indicates the control action for the process. Generally, fuzzy rules of the Mamdani type have the structure shown in the Eq. (2), which are the most used for the simulation and implementation of a fuzzy controller. The knowledge of a person can be used, or the simulation of the process can be carried out to determine the fuzzy rules. The inference stage determines the fuzzy sets that will be used in the defuzzification stage. Finally, **Table 1** shows the fuzzy rules that relate the membership values of the fuzzification stage used to determine the inference matrix and with it, the value of the tip [24, 25].

$$\text{IF x is A AND y is B THEN z is C} \tag{2}$$

		Service				
Food		**VB**	**BD**	**RG**	**GD**	**VG**
	VB	VB	VB	BD	BD	RG
	BD	VB	BD	BD	RG	RG
	RG	BD	BD	RG	RG	GD
	GD	BD	RG	RG	GD	GD
	VG	RG	RG	GD	GD	EX

Table 1.
Knowledge base or inference matrix of the fuzzy rules to determine the tip.

Where: x and y are the input variables, z is the output variable, A and B are fuzzy sets of the input variables, and C is a fuzzy set of the output variable.

The algorithm used to implement fuzzy rules in MATLAB™ Script and Arduino UNO, Arduino DUE, and Nexys 4™ boards is described below.

Eq. (2) is used to declare fuzzy rules. Setfd and Setsv were defined in the fuzzification stage, and Settp contains the result of the fuzzy rule (output fuzzy set).

% MATLAB Script	*// Arduino UNO / DUE*	*– Nexys 4*
if Setfd==VB && Setsv==VB	if (Setfd==VB && Setsv==VB) {	if (Setfd=VB and Setsv=VB) then
Settp=VB;	Settp = VB;	Settp := VG;
end	}	end if;
if Setfd==VG && Setsv==VG	if (Setfd==VG && Setsv==VG) {	if (Setfd=VG and Setsv=VG) then
Settp =EX;	Settp = EX;	Settp := EX;
end	}	end if;

Also, the inference stage must define a membership function μ(x) for the output fuzzy set or result of a fuzzy rule. Generally, the inference method of the Mamdani type (min-max) is used for the implementation of a fuzzy controller, which uses Eq. (3), which selects the membership function of the input variable with the minimum value. In this case, the IF-THEN control statement is used to determine the membership function with the minimum value, which reduces the computational load and allows the inference stage to be implemented on different platforms that cannot use the min (a, b) operation. Finally, there are fuzzy rules that have the same result (output fuzzy set), which implies that multiple membership functions can be associated with an output fuzzy set.

$$\mu_C(x) = \min\{\mu_A(x), \mu_B(x), \dots, \mu_M(x)\} \tag{3}$$

where: $\mu_A(x), \mu_B(x), \dots, \mu_M(x)$ are the membership functions of the input variables, and $\mu_C(x)$ is the membership function for a fuzzy set of the output variable.

The algorithm used to implement the inference method in MATLAB™ Script and Arduino UNO, Arduino DUE, and Nexys 4™ boards is described below.

MfOut is the membership function of the output fuzzy set (result of a fuzzy rule), and the variables "Mf1" and "Mf2" were defined in the fuzzification stage.

% MATLAB Script	*// Arduino UNO / DUE*	*– Nexys 4*
if Mf1 > Mf2	if (Mf1 > Mf2) {	if (Mf1 > Mf2) then
MfOut = Mf2; % Mf2<Mf1	MfOut = Mf2; // mf2<mf1	MfOut := Mf2; – mf2 < mf1
else	else	else
MfOut = Mf1; % Mf1<Mf2	MfOut = Mf1; // mf1<mf2	MfOut := Mf1; – mf1<mf2
end	}	end if;

2.5 Aggregation stage

The aggregation stage is used to determine and obtain the membership functions μ(x) of the output fuzzy sets, which will be used in the defuzzification stage. As mentioned above, the inference stage can associate multiple membership functions to an output fuzzy set. Therefore, the aggregation stage uses Eq. (4) to select the highest value of the membership function of the aggregation stage. In this case, IF-THEN conditional statements were used to compare the multiple membership functions and obtain the highest value of the membership function. This option reduces the computational load and allows the aggregation stage to be implemented on different platforms, which cannot use the *max* (a, b) operation [26, 27].

$$\mu_C(x) = \max(\mu_{C1}(x), \mu_{C2}(x), \dots, \mu_{CM}(x)) \tag{4}$$

where: $\mu_C(x)$ is the membership function of output fuzzy set, which will be used for defuzzification, and $\mu_{C1}(x)$, $\mu_{C2}(x)$, ..., $\mu_{CM}(x)$ are the membership functions defined for the output fuzzy set.

The algorithm used to implement the aggregation stage in the MATLAB ™ Script and on the Arduino UNO, Arduino DUE, and Nexys 4™ boards is described below.

Mftip is the membership function of the output fuzzy set, which will be used in the defuzzification, and MfOut1, ..., MfOut3 are membership functions, which were defined for an output fuzzy set.

% MATLAB Script	*// ArduinoUNO / DUE*	*– Nexys 4*
if MfOut1 >= MfOut2	if (MfOut1 >= MfOut2) {	if (MfOut1 >= MfOut2) then
Mftip = MfOut1;	Mftip = MfOut1;	Mftip := MfOut1;
else	else	else
Mftip = MfOut2;	Mftip = MfOut2;	Mftip := MfOut2;
end	}	end if;
if Mftip >= MfOut3	if (Mftip >= MfOut3) {	if (Mftip >= MfOut3) then
Mftip = Mftip;	Mftip = Mftip;	Mftip = Mftip;
else	else	else
Mftip = MfOut3;	Mftip = MfOut3;	Mftip = MfOut3;
End	}	end if;

2.6 Defuzzification stage

The last stage of the fuzzy controller is defuzzification, which is used to determine a numerical value, which represents the output fuzzy sets. In this work, the defuzzification value is determined using Eq. (5), which represents the centroid method. The centroid method is used, since this method only requires the basic operations of addition, subtraction, multiplication, and division for its implementation in software or hardware. The defuzzification value represents the rigid value of the output controller, in this case, the value of the tip of the food establishment. Finally, the defuzzification value represents the control action that must be performed in a process [20, 28, 29].

$$\text{defuzzification} = (x_1 {}^* \mu(x_1) + x_2 {}^* \mu(x_2) + \dots + x_n {}^* \mu(x_n))/(\mu(x_1) + \mu(x_2) + \dots + \mu(x_n)) \tag{5}$$

where: x_1, x_2, ..., x_n are values of the output variable ("tip"), which are found within the fuzzy sets obtained for defuzzification, and $\mu(x_1)$, $\mu(x_2)$, ..., $\mu(x_n)$ are the membership functions of x_1, x_2, ..., x_n.

The algorithm used to implement the defuzzification stage in the MATLAB Script and on the Arduino UNO, Arduino DUE, and Nexys 4™ boards is described below.

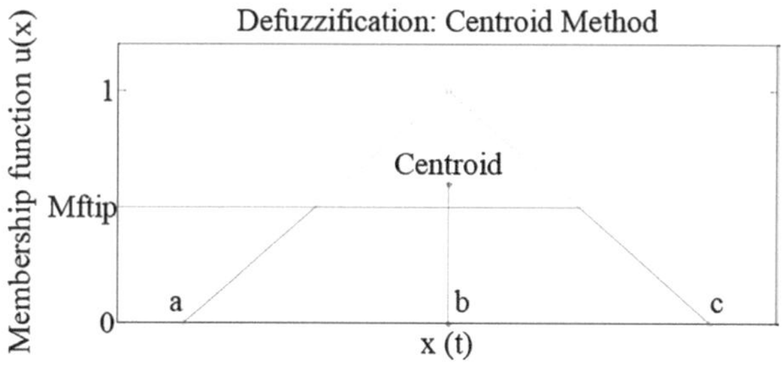

Figure 5.
Defuzzification using the centroid method.

Figure 5 shows the fuzzy set used to show the defuzzification process. Eq. (5) is used to determine the defuzzification value. Mftip is the membership function defined in the aggregation stage, mfx is the membership function of x, num1 and den1 are the numerator and denominator of Eq. (5), respectively.

% MATLAB Script	// Arduino UNO / DUE	– *Nexys 4™*
Tpaso=(c - a) / 1000; x=a;	Tpaso=(c - a) / 1000; x=a;	Tpaso=(c - a) / 10; x = a;
for k=1:1:1000	for (k=1; k<=1000; k++){	for i in 0 to 10 loop
x = x + Tpaso;	x = x + Tpaso;	if (x >= a and x <= b) then
if x >= a && x <= b	if (x >= a && x <= b){	mux := ((x-a)*(100))/(b-a);
mfx = (x - a) / (b - a);	mfx = (x - a) / (b - a);	if (mfx >= Mftip) then
if mfx >= Mftip	if (mfx >= Mftip){	mfx := Mftip;
mfx = Mftip;	mfx = Mftip;	end if; end if;
end end	}}	if (x > b and x <= c) then
if x > b && x <= c	if (x > b && x <= c){	mux := ((c-x)*(100))/(c-b);
mfx=(c - x)/(c - b);	mfx=(c - x)/(c - b);	if (mfx >= Mftip) then
if mfx >= Mftip	if (mfx >= Mftip){	mfx := Mftip;
mux= Mftip;	mux= Mftip;	end if; end if;
end end	}}	num1 := num1+(x*mfx);
num1=num1+(x*mfx);	num1=num1+(x*mfx);	den1 := den1 + mfx;
den1=den1+mfx; end	den1=den1+mfx; }	end loop;
% Defuzzification result	// Defuzzification result	– Defuzzification result
dzz=num1/den1;	dzz=num1/den1;	dzz := num1 / den1;

2.7 Procedure for the implementation of a fuzzy controller in the process

As mentioned above, a fuzzy controller is used to control some variables of a process; therefore, the controlled variable, the type of sensor to measure the controlled variable, and the type of actuator (DC motor, stepper motor, fan, heater, etc.) required to adjust the process must be defined. Additionally, the resolution of the sensor (8 bits, 10 bits, 12 bits, etc.), the sampling rate of the sensor, the sensor operating range, and the type of signal (the number of steps of a stepper motor, minimum and maximum speed of a fan, PWM signal, etc.) required to control the movement of the actuator must be determined to control a process using a fuzzy controller. In this case, the sensor operating range is used to define the characteristics of the input variables of the fuzzy controller and the working range of the actuator is used to define the characteristics of the output variable of the controller [30]. **Figure 6** shows a block diagram of a control system for a process, which uses a generic fuzzy controller. The controller uses the error e(t) and the derivative of the error d(e(t))/dt as input variables, the output variable is a control signal for the actuator (mv_{act}). Finally, a fuzzy controller does not allow an overshoot to be generated in the system

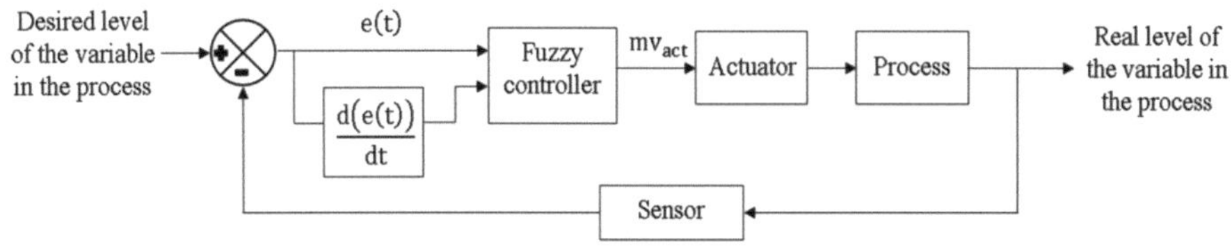

Figure 6.
Block diagram of the control of a process using a fuzzy controller.

response like a classical controller. This is because a fuzzy controller relates the input variables to the output variable through fuzzy rules of the IF-THEN type. So the output response does not have oscillations in the output, and only the settling time and rise time of the system response are short.

3. Experimental and simulation results

The analysis, design, and simulation of the programming algorithm for the fuzzy controller were carried out in a MATLAB™ Script, which was used to perform an analysis of its operation. **Figure 7** shows the simulation of the fuzzy controller in the MATLAB™ Script, **Figure 7(a)** shows the fuzzification of the "food" variable with a score of 15, **Figure 7(b)** shows the fuzzification of the "service" variable with a score of 35, and **Figure 7(c)** shows a defuzzification value or tip value of $31.61. Fuzzy Logic Toolbox™ was used to determine the accuracy of the fuzzy controller, which was implemented using the proposed methodology in the different platforms mentioned above. Therefore, the results of Fuzzy Logic Toolbox™ are considered as the ideal results or correct results. The results of the fuzzy controller in the MATLAB™ Script (MS), Fuzzy Logic Toolbox™ (FLT), the Arduino UNO board (AUNO), the Arduino DUE board (ADUE), and the Nexys 4™ board (NX4) are shown in **Table 2**. The Mean Square Error (MSE) is used to as an analysis of the accuracy of the proposed

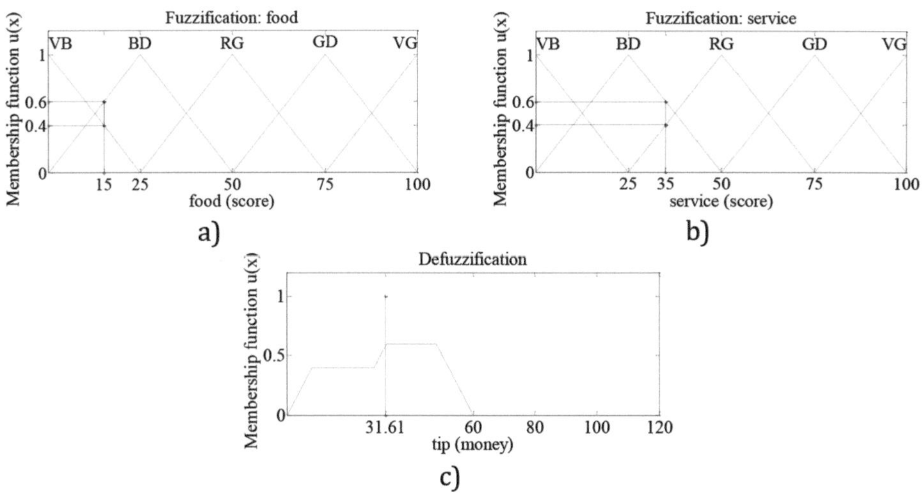

Figure 7.
Result of (a) fuzzification of the food variable, (b) fuzzification of the service variable, and (c) defuzzification value or tip value.

Input variables		Defuzzification (tip)				Mean Square Error (MSE)		
Food	Service	FLT	MS	AUNO/ ADUE	NX4	MS	AUNO/ADUE	NX4
0	0	20.0	20.0000	20.00	20	0.0000	0.0000	0.00
10	20	28.4	28.3873	28.38	28	0.0000	0.0002	0.08
20	5	24.8	24.8278	24.83	25	0.0003	0.0004	0.02
30	40	45.4	45.3849	45.38	45	0.0001	0.0002	0.08
40	80	65.4	65.3849	65.38	65	0.0001	0.0002	0.08
50	25	40.0	40.0000	40.00	40	0.0000	0.0000	0.00
60	5	45.4	45.3849	45.38	45	0.0001	0.0002	0.08
70	95	75.2	75.1726	75.17	75	0.0003	0.0004	0.02
80	10	48.4	48.3873	48.39	48	0.0000	0.0000	0.08
90	45	71.6	71.6131	71.61	72	0.0000	0.0000	0.08
100	100	100.0	100.0000	100.00	100	0.0000	0.0000	0.00

Table 2.
Experimental results and simulation results of fuzzy controller.

methodology, which is summarized in **Table 2**. It can be seen that the fuzzy controller implemented in the Arduino boards, Nexys 4™, and the MATLAB™ Script generate almost the same MSE in all the examples. Also, the results show that the fuzzy controller can estimate the optimal parameters and compensate the uncertainties and nonlinearity of the system. As we can see, the error is minimal in the different platforms. This is very important for the research presented here, which focuses on the retention of experience and its subsequent instead of the mathematical model and nonlinearities found in the system.

The MSE, which measures the error between two datasets, was used to determine the accuracy of the proposed fuzzy controller. Eq. (6) was used to determine the MSE, and the values of 0.0326, 0.0643, and 0.1125 were obtained for the Arduino UNO, Arduino DUE, and Nexys 4™ boards, respectively. A high degree of accuracy is obtained on the MATLAB™ and Arduino platforms, since their programming language allows the use of a wide variety of variable types (integers, floating point, bit, byte, etc.) and control statements (IF-THEN, FOR, WHILE, etc.). Finally, the precision of the Nexys 4™ board is lower, since this board does not allow the use of floating-point numbers or real numbers.

$$\mathrm{MSE} = \frac{1}{n}\sum_{i=1}^{n} (dn_i - y_i)^2 \tag{6}$$

where n is the data number, d_n is the desired value, and y_n is the system result.

An analysis of the control surface or controllability of the process is carried out, which is obtained from the fuzzy controller results. The control surface shows the form, in which the process control will be performed. The control surface shows the mapping of the input and output variables of the controller. The control surfaces for the MATLAB™ Script, Fuzzy Logic Toolbox™, Arduino UNO board, Arduino DUE board, and Nexys 4™ board are shown in **Figure 8**. We can see when achieving a monotonic curve as **Figure 8** has, means that do not exist large changes and for this,

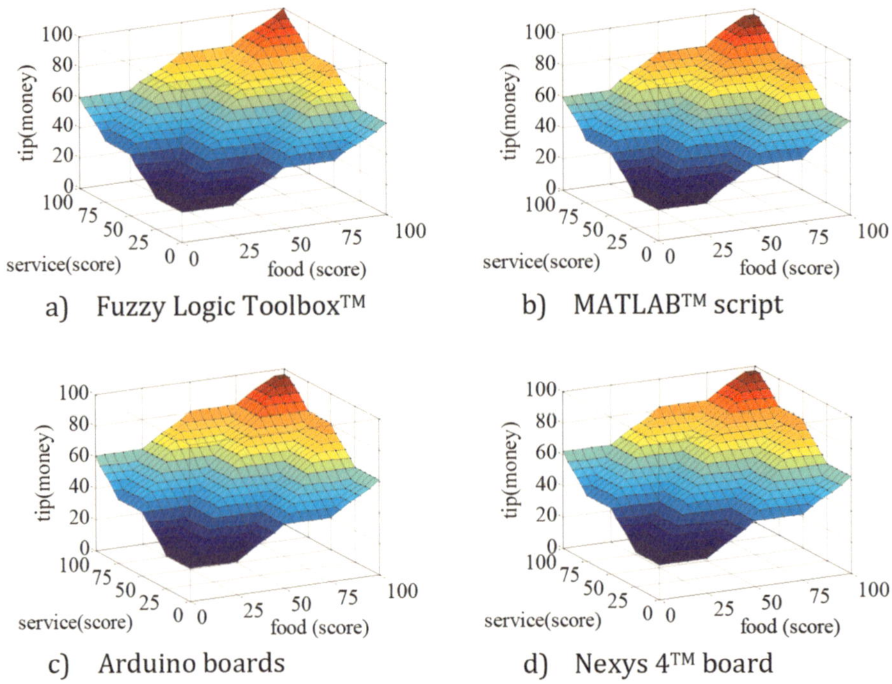

Figure 8.
Control surface of (a) Fuzzy Logic Toolbox™, (b) MATLAB™ script, (c) Arduino boards, and (d) Nexys 4™ board.

control action is smooth that means that the movement of the actuators works without stress, since an unsmooth control action can damage the actuators in the process. The control surface shows all the possible results of the controller; that is, the control surface shows all the results of the variable "tip" for all possible combinations of the variables "food" and "service."

3.1 Characteristics of the implementation of the fuzzy controller

The amount of memory and the processing time are the most important aspects when implementing a programming algorithm on a hardware board. Therefore, the amount of memory and the processing time of the fuzzy controller should be analyzed. The fuzzy controller, which was implemented on the Arduino UNO board, has a processing time of 117 ms and uses 40% of the board's memory. The fuzzy controller, which was implemented on the Arduino DUE board, has a processing time of 21.275 ms and uses 5% of the board's memory. The fuzzy controller, which was implemented on the Nexys 4™ board, has a processing time of 17.871 milliseconds and uses 40% of the board's memory.

3.2 Comparison between the methodology fuzzy controller proposed with other works

As a comparison, the analysis of the accuracy of the fuzzy controller proposed with different research works was taken into account, in which a fuzzy controller was used. In this case, Refs. [31, 32] was used to compare the efficiency. Arduino Mega 2560 board and the Arduino UNO board are compared. **Tables 3-5** show the results of the fuzzy controller and the MSE of the data. As can be seen, the fuzzy controller using the proposed methodology has a higher precision than the fuzzy controller taking in consideration.

On the other hand, there are research works that use a computer to implement a fuzzy controller and a hardware board as an interface to interconnect the environment (physical variables) with the computer [33, 34]. This action can increase the cost of the system and make it difficult to implement it in a process. **Table 6** shows the results of a fuzzy controller, which is implemented in MATLAB™ and used for the prediction of GSM tissue and wrinkle recovery angle of laser-engraved denim [35]. Additionally, the results of the fuzzy controller implemented using the proposed methodology and the MSE are shown. As can be seen, a high degree of accuracy can be obtained in a process using the methodology proposed in this work.

Light intensity	Brightness (LED)			Mean Square Error	
	Fuzzy Logic Toolbox™	Fuzzy controller of the Greenhouse (Arduino Mega)	Proposed fuzzy controller (Arduino uno)	Arduino MEGA	Arduino UNO
523	450	450	450.00	0.0000	0.0000
498	455	455.85	454.93	0.3612	0.0024
78	850	850	850.00	0.0000	0.0000
57	850	850	850.00	0.0000	0.0000
218	810	808.5	809.60	1.1250	0.0800
213	820	818.4	820.00	1.2800	0.0000
693	267	269.2	266.70	2.4200	0.0450
689	276	278.9	275.64	4.2050	0.0648
801	250	250	250.00	0.0000	0.0000
688	278	281.2	277.80	5.1200	0.0200

Table 3.
Test of the light intensity of an LED using a diffuse driver.

PH	Moisture (%)	Experimental results of the process	Fuzzy controller for fertilization (MATLAB)	Proposed fuzzy controller (Arduino uno)	Mean Square Error	
					MATLAB	Arduino UNO
5.3	43	0 ml	0 ml	0 ml	0.0000	0.0000
6.89	71	18 ml	18.35 ml	18.37	0.0612	0.0684
4.01	57	0 ml	0 ml	0 ml	0.0000	0.0000
7.15	59	23 ml	23.23 ml	23.31 ml	0.0264	0.0480

Table 4.
Results of the acid solution for the process using a fuzzy controller.

PH	Moisture (%)	Experimental results of the process	Fuzzy controller for fertilization (MATLAB)	Proposed fuzzy controller (Arduino uno)	Mean Square Error	
					MATLAB	Arduino UNO
5.3	43	55 ml	54.68 ml	54.53 ml	0.0512	0.1104
6.89	71	0 ml	0 ml	0 ml	0.0000	0.0000
4.01	57	25 ml	25.49 ml	25.64 ml	0.1200	0.2048
7.15	59	0 ml	0 ml	0 ml	0.0000	0.0000

Table 5.
Results of the neutral solution for the process using a fuzzy controller.

Dots Per Inch	Pixel Time	Fabric weight (GSM)	Fuzzy controller to predict the GSM	Proposed fuzzy controller (Arduino uno)	Recovery Angle (Degree)	Fuzzy controller to predict the angle of recovery (Degree)	Proposed fuzzy controller (Arduino uno)
15	100	439	429	428.5105	77	75	75.00
15	150	411	409	408.8334	71	70.8	70.83
15	200	385	379	378.3335	66	64.5	64.54
20	100	400	409	408.8334	69	70.8	70.83
20	150	375	379	378.3335	65	64.5	64.54
20	200	354	348	348.1668	60	58.3	58.38
25	100	372	379	378.3335	62	64.5	64.54
25	150	345	348	348.1668	58	58.3	58.38
25	200	318	328	328.1575	52	54	54.23

Table 6.
Fuzzy controller to predict the strength values of laser-engraved denim seams.

4. Conclusions

In this work, a methodology was proposed to implement a fuzzy controller of the Mamdani type on different platforms (software or hardware). The methodology is based on the IF-THEN and FOR control statements, and mathematical operations such as addition, subtraction, multiplication, and division, which are used to establish each of the stages of a fuzzy controller (fuzzification, fuzzy rules, or defuzzification). The key ideas explored are the use of programming languages such as VHDL, C++, and MATLAB™. The operation of the fuzzy controller based on the proposed methodology is tested using different examples, which are compared with the Fuzzy Logic Toolbox™ (MATLAB™ tool). The results show that the fuzzy controller methodology can improve the convergence of the system and improve the error properties. In addition, the proposed methodology reduces the computational difficulty and the computational load. Also, the procedure to implement the fuzzy controller in a process is described. Compared to other works, the greatest contribution of this work is to describe the elements to program and implement each of the stages of a fuzzy controller in software or hardware. This will allow estimating the optimal parameters, uncertainties, and nonlinearities of a dynamic system to control a system without the mathematical model. Finally, simulations were shown to verify that the proposed methodology, and the theoretical and experimental results are valid.

Acknowledgements

This study was funded by Consejo Mexiquense de Ciencia y Tecnologia (COMECYT).

Author details

Jesus de la Cruz-Alejo*, Hugo Beatriz-Cuellar, Agustin Mora-Ortega and Maria Belem Arce-Vazquez
Tecnológico de Estudios Superiores de Ecatepec, Estado de Mexico, Mexico

*Address all correspondence to: jdelacruz@tese.edu.mx

References

[1] Zou Y, Yan F, Wang X, Zhang J. An efficient fuzzy logic control algorithm for photovoltaic maximum power point tracking under partial shading condition. Journal of the Franklin Institute. 2020; **357**(6):3135-3149. DOI: 10.1016/j.jfrank lin.2019.07.015

[2] Subramanian V, Indragandhi V, Kuppusamy R, Teekaraman Y. Modeling and analysis of PV system with fuzzy logic MPPT technique for a DC microgrid under variable atmospheric conditions. Electronics (Basel). 2021; **10**(20):2541. DOI: 10.3390/electronics 10202541

[3] Hassan T-U, Abbassi R, Jerbi H, Mehmood K, Tahir MF, Cheema KM, et al. A novel algorithm for MPPT of an isolated PV system using push pull converter with fuzzy logic controller. Energies. 2020; **13**(15):4007. DOI: 10.3390/en131 54007

[4] Farajdadian S, Hosseini SMH. Design of an optimal fuzzy controller to obtain maximum power in solar power generation system. Solar Energy. 2019; **182**:161-178. DOI: 10.1016/j.solener. 2019.02.051

[5] Loukil K, Abbes H, Abid H, Abid M, Toumi A. Design and implementation of reconfigurable MPPT fuzzy controller for photovoltaic systems. Ain Shams Engineering Journal. 2020;**11**(2): 319-328. DOI: 10.1016/j.asej.2019. 10.002

[6] Muñoz Villacorta R, Oscco Agüero C, Andrade-Arenas L. Implementation of an intelligent system for the diagnosis and treatment of venereal diseases. International Journal of Online Engineering. 2022;**18**(11):58-76. DOI: 10.3991/ijoe.v18i11.32329

[7] Aamir KM, Sarfraz L, Ramzan M, Bilal M, Shafi J, Attique M. A fuzzy rule-based system for classification of diabetes. Sensors (Basel). 2021;**21**(23): 8095. DOI: 10.3390/s21238095

[8] Zulkarnain AF, Sari Y, Rakhmadani R. Monitoring system for early detection of fire in wetlands based internet of things (IoT) using fuzzy methods. IOP Conference Series: Materials Science and Engineering. 2021; **1115**(1):012007. DOI: 10.1088/ 1757-899x/1115/1/012007

[9] Soliman MA, Hasanien HM, Azazi HZ, El-Kholy EE, Mahmoud SA. An adaptive fuzzy logic control strategy for performance enhancement of a grid-connected PMSG-based wind turbine. IEEE Transactions on Industrial Informatics. 2019;**15**(6):3163-3173. DOI: 10.1109/tii.2018.2875922

[10] Ganthia BP, Barik SK. Fault analysis of PI and fuzzy-logic-controlled DFIG-based grid-connected wind energy conversion system. Journal of The Institution of Engineers (India): Series B. 2022;**103**(2):415-437. DOI: 10.1007/s40031-021-00664-9

[11] Ngo Q-V, Yi C, Nguyen T-T. The maximum power point tracking based-control system for small-scale wind turbine using fuzzy logic. International Journal of Electrical and Computer Engineering (IJECE). 2020;**10**(4):3927. DOI: 10.11591/ijece.v10i4.pp3927-3935

[12] Murillo-Yarce D, Munoz J, Restrepo C. Mamdani type PI-fuzzy controller in a boost converter. IEEE International Conference on Industrial Technology (ICIT). 2020:487-492. DOI: 10.1109/ ICIT45562.2020.9067257

[13] Riahi J, Vergura S, Mezghani D, Mami A. Intelligent control of the

microclimate of an agricultural greenhouse powered by a supporting PV system. Applied Sciences (Basel). 2020;**10**(4):1350. DOI: 10.3390/app10041350

[14] Sarmiento G. Arduino Curso Práctico: Manual Práctico. North Charleston, SC, United States of America: Createspace Independent Publishing Platform; 2015

[15] McRoberts M. Beginning Arduino. 2nd ed. New York, NY, United States of America: APRESS; 2013

[16] Maxinez DG, Alcalá Jara J. VHDL. El arte de programar sistemas digitales. 1st ed. CECSA; 2002. p. 368

[17] Nexys 4TM FPGA Board Reference Manual [Internet]. Digilent.com. 2016. Available from: https://digilent.com/reference/_media/nexys:nexys4:nexys4_rm.pdf

[18] Kalechman M. Practical MATLAB Applications for Engineers. Londres, Inglaterra: CRC Press; 2018

[19] Knight A. Basics of MATLAB and Beyond. 1st ed. Chapman and Hall/CRC; 2019. p. 216. DOI: 10.1201/9780429186882

[20] Michels K, Klawonn F, Kruse R, Nurnberger A. Fuzzy Control: Fundamentals, Stability and Design of Fuzzy Controllers. Berlín, Germany: Springer; 2010

[21] Blokdyk G. Fuzzy Logic: Beyond the Basics. North Charleston, SC, United States of America: Createspace Independent Publishing Platform; 2017

[22] Santos W, editor. Fuzzy Control Systems: Design, Analysis & Performance Evaluation. Nova Science: Hauppauge, NY, United States of America; 2017

[23] Ebrahimnejad A, Verdegay JL. Fuzzy Set Theory. En: Fuzzy Sets-Based Methods and Techniques for Modern Analytics. Cham: Springer International Publishing; 2018. pp. 1-27

[24] Fuzzy logic projects with matlab. En: Introduction to Fuzzy Logic using MATLAB. Berlin, Heidelberg: Springer Berlin Heidelberg; 2007. pp. 369–408

[25] Zadeh LA, Aliev RA. Fuzzy Logic Theory and Applications: Part I and Part II: Part I and Part II. Singapur, Singapur: World Scientific Publishing; 2019

[26] Ross TJ. Fuzzy Logic with Engineering Applications. 4a ed. Nashville, TN, United States of America: John Wiley & Sons; 2016

[27] Gerla G. Fuzzy Logic: Mathematical Tools for Approximate Reasoning. Dordrecht, Netherlands: Springer; 2010

[28] Kovacic Z, Bogdan S. Fuzzy Controller Design: Theory and Applications. 1st ed. CRC Press; 2019. p. 416. DOI: 10.1201/9781420026504

[29] Pourabdollah A. Fuzzy Number Value or Defuzzified Value; Which One Does It Better?. IEEE International Conference on Fuzzy Systems (FUZZ-IEEE). 2020:1-6. DOI: 10.1109/FUZZ48607.2020.9177533

[30] Rodriguez Lopez ML. Tendencias en Instrumentación y Control de Procesos. Encuentro Sennova del Oriente Antioqueño. 2021. DOI: 10.23850/22565035.2953

[31] Kurniawan D, Witanti A. Prototype of Control and Monitor System with Fuzzy Logic Method for Smart Greenhouse. Indonesian Journal of Information Systems.2021;**3**(2):116-127. DOI: 10.24002/ijis.v3i2.4067

[32] Penzol N, Adnan R. Design of an Internet of Things (Iot) Based Smart Irrigation and Fertilization System Using Fuzzy Logic for Chili Plant. Shah Alam, Malaysia: IEEE International Conference on Automatic Control and Intelligent Systems (I2CACIS 2020); 2020. DOI: 10.1109/I2CACIS49202.2020.9140199

[33] Hassan T, Abbassi R, Jerbi H, Mehmood K, Tahir MF, Cheema KM. A Novel Algorithm for MPPT of an Isolated PV System Using Push Pull Converter with Fuzzy Logic Controller. Energies. 2020;**13**(15):4007. DOI: 10.3390/en13154007

[34] Dehghani M, Taghipour M, B. Gharehpetian G, Abedi M. Optimized Fuzzy Controller for MPPT of Grid-connected PV Systems in Rapidly Changing Atmospheric Conditions. Journal of Modern Power Systems and Clean Energy. 2021;**9**(2):376-383. DOI: 10.35833/MPCE.2019.000086

[35] Sarkar J, Mondal M, Khalil E. Predicting fabric GSM and crease recovery angle of laser engraved denim by fuzzy logic analysis. Journal of Engineering and Applied Science. 2020; **4**(1):52-64

Chapter 8

Performance Improvement for Fighter Aircraft Using Fuzzy Switching LQI Controller

Emre Kemer, Hasan Başak and Hayri Baytan Özmen

Abstract

In this work, a switching linear quadratic integral (LQI) controller based on fuzzy logic is designed for the load-factor tracking problem of high-performance aircraft referred to as the Aero-Data Model in Research Environment (ADMIRE). ADMIRE is a new generation aircraft and has a wide flight operation envelope in terms of altitude and speed. Hence, it is difficult to design a flight controller to achieve a high tracking performance. First, the LQI controller is selected due to good tracking performance and robustness in the model dynamics. Combining switching LQI controller and fuzzy logic improves the transient performance of the closed-loop switched system. The results obtained with the fuzzy switching controller have been compared with the conventional LQI and the switching LQI in terms of robust demand tracking. The simulation results have demonstrated that the fuzzy switching controller is superior to the conventional LQI and switching LQI controllers due to better transient performance and robust stability.

Keywords: fuzzy logic, switching control, LQI, load-factor tracking, fighter aircraft

1. Introduction

Conventional aircrafts have aileron, elevator, and rudder control effectors. Flight control systems are generally developed using one control effector for each rotational degree of freedom. The aileron is utilized to obtain a roll motion, a pitch motion is obtained by using the elevator, and the rudder effector controls the yaw motion of the aircraft. The control problem is the determination of the deflections of control effectors that produce the desired motion specified by a flight controller that transfers the pilot's command given by a control stick. Three control effectors can generate desired motions. However, modern aircrafts have more control effectors than conventional aircrafts [1]. The design of reliable flight control systems is difficult for modern aircrafts because these aircrafts are becoming more complicated. Also, the performance of flight control systems must be very high and the stability of the aircraft has demanded the development of different control systems [2]. In recent years, linear control systems have been developed assuming that flight dynamics are linear time-invariant about the operation points and the longitudinal dynamics are decoupled from lateral ones. Zhang et al. [3] proposed a mixed H_2/H_∞ flight controller using enhanced linear matrix inequality, which stabilizes the aircraft system in case of actuator loss. A gain scheduled linear quadratic regulator method is designed in [4] for vehicle dynamics where the flight period is divided into different intervals because flight condition varies during the flight. A proportional-integral-derivative (PID) flight control system is investigated in ref. [5] whose performance is not satisfactory due to uncertainties and nonlinearities of vehicle dynamics. A flight controller law is designed based on optimal control theory in ref. [6] ensuring the reliability of aircraft for pilot's commands in case of all operating conditions. A resilient linear controller is proposed by Bouvier et al. [7] for the dynamic of aircraft in the presence of a loss of control authority. Offline reference regulators and robust control allocation flight controllers were developed in ref. [8] for aerodynamic nonlinearities and parametric uncertainties. Besides, nonlinear controller methods have been proposed by researchers. For example, a nonlinear dynamic inversion control law is proposed by Da Costa et al. [9] where the nonlinear dynamics are transformed into linear dynamics using state or output feedback assuming timescale separation between attitude and altitude rates. Nonlinear dynamic inversion controllers require precise knowledge of all nonlinearities that is not possible for modern fighter aircraft [10]. Sliding mode differentiator [11], disturbance observer-based sliding mode control [12], and disturbance observer-based dynamic surface controller [13] are developed considering nonlinearities and external disturbances.

A backstepping control based on fuzzy logic system is designed in ref. [14] for vehicle dynamics with state constraints and actuator fault. A fuzzy tracking controller [15] was proposed to satisfy the properties of disturbance rejection in aircraft vehicles. Takagi-Sugeno fuzzy robust controller was developed by Luan et al. [16] for the problem of part transportation. An adaptive fuzzy controller [17] was designed for a vehicle dynamic with input saturation.

In this chapter, we develop a control approach based on a switching control with a fuzzy logic rule, which is evaluated in a nonlinear ADMIRE aircraft model. Combined switching control with fuzzy logic has better tracking performance and strong robustness for the nonlinear model of ADMIRE aircraft.

2. The ADMIRE aircraft model

The Aeronautical Research Institute of Sweden developed the ADMIRE model using the generic aero-data model with dynamic models of an engine, actuators, atmosphere, and sensors. The ADMIRE model has 12 states but generally, these states were reduced to simply nonlinear dynamics of the system. The short-period longitudinal flight dynamics governing the ADMIRE benchmark model are given as follows [18–20]:

$$\begin{vmatrix} \dot{\alpha} \\ \dot{q} \end{vmatrix} = \begin{bmatrix} Z_\alpha & Z_q \\ M_\alpha & M_q \end{bmatrix} \begin{bmatrix} \alpha \\ q \end{bmatrix} + \begin{bmatrix} Z_{\delta_e} & Z_{t_{ss}} \\ M_{\delta_e} & M_{t_{ss}} \end{bmatrix} \begin{bmatrix} \delta_e \\ t_{ss} \end{bmatrix}$$

$$\mathrm{n_z} = \begin{bmatrix} \mathrm{n_{z\alpha}} & \mathrm{n_{zq}} \end{bmatrix} \begin{bmatrix} \alpha \\ \mathrm{q} \end{bmatrix} + \begin{bmatrix} \mathrm{n_{z\alpha}} & \mathrm{n_{zq}} \end{bmatrix} \begin{bmatrix} \delta_\mathrm{e} \\ \mathrm{t_{ss}} \end{bmatrix} \tag{1}$$

where state variables $\boldsymbol{\alpha}$ and $\boldsymbol{q}$ are the angles of attack and the Euler pitch rate, respectively. The control inputs are the elevator angle, $\boldsymbol{\delta_e}$ and throttle setting, $\boldsymbol{t_{ss}}$, respectively, and the output variable is load-factor, $\boldsymbol{n_z}$ (**Figure 1**).

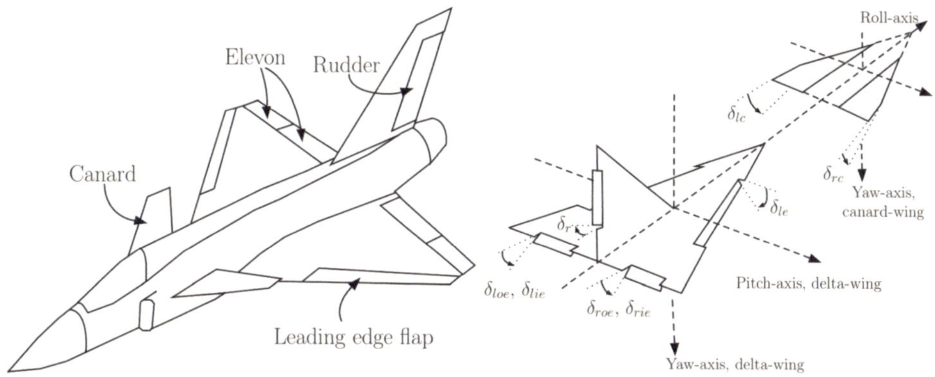

Figure 1.
ADMIRE aircraft model and control surfaces.

3. Fuzzy switching control development

In this section, a fuzzy switching control will be developed for the ADMIRE fighter aircraft. **Figure 2** illustrates a schematic of the control structure. Here, linear quadratic integral (LQI) control computes an optimal state feedback gain for the regulating closed-loop system. The control law consists of the solution of the Riccati equation in the linear-quadratic regulatory framework with the integral of the output variable. The linearized dynamics of the aircraft at a trim condition with state-space realization are given as:

$$\dot{x}(t) = Ax(t) + Bu(t)$$
$$y(t) = Cx(t) + Du(t) \tag{2}$$

The objective of the LQI control is to find the state feedback control law, such as

$$u(t) = -K[x(t)e_I(t)]^T \tag{3}$$

where $\boldsymbol{K}$ is the feedback gain matrix, and $\boldsymbol{e_I(t)}$ is the integral state for the output variable. The optimal feedback law minimizes the quadratic performance index.

$$J = \int_0^{\infty} \left(x^T Q x + x^T R\, u\right) dt \tag{4}$$

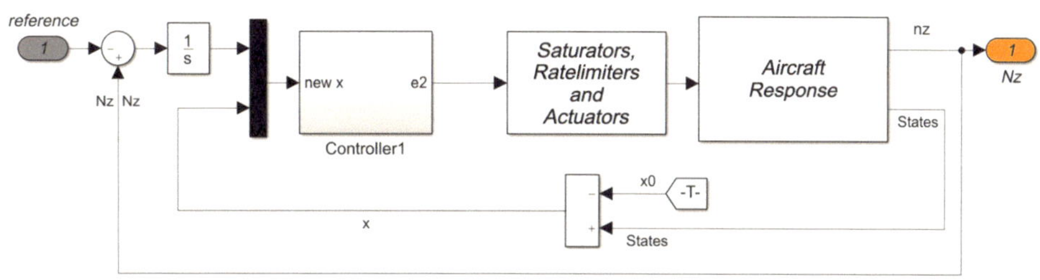

Figure 2.
Schematic of the control structure.

In which $\boldsymbol{Q}$ is a positive semi-definite weight matrix, and $\boldsymbol{R}$ is a positive-definite weight matrix.

Then, this control law guarantees that the output $\boldsymbol{y}(\boldsymbol{t})$ tracks the demand signal $\boldsymbol{r}(\boldsymbol{t})$. In fact, $\boldsymbol{e_I}(\boldsymbol{t})$ is

$$e_I(t) = \int_0^t (r(\tau) - y(\tau))d\tau \tag{5}$$

The state-space presentation of augmented dynamic is written as:

$$\begin{vmatrix} \dot{x}(t) \\ \dot{e}_I(t) \end{vmatrix} = \begin{vmatrix} A(t) & 0 \\ -C(t) & 0 \end{vmatrix} \begin{vmatrix} x(t) \\ e_I(t) \end{vmatrix} + \begin{vmatrix} B(t) \\ -D(t) \end{vmatrix} u(t) \tag{6}$$

To cover the flight envelope, the flight envelope is divided into some cells. Augmented switched state-space model is given as:

$$\begin{vmatrix} \dot{x}_{\sigma(t)}(t) \\ \dot{e}_{I\sigma(t)}(t) \end{vmatrix} = \begin{vmatrix} A_{\sigma(t)}(t) & 0 \\ -C_{\sigma(t)}(t) & 0 \end{vmatrix} \begin{vmatrix} x_{\sigma(t)}(t) \\ e_{I\sigma(t)}(t) \end{vmatrix} + \begin{vmatrix} B_{\sigma(t)}(t) \\ -D_{\sigma(t)}(t) \end{vmatrix} u(t) \tag{7}$$

The system matrices of Eq. (7) are rewritten as:

$$\begin{vmatrix} A_{\sigma(t)} & B_{\sigma(t)} \\ C_{\sigma(t)} & D_{\sigma(t)} \end{vmatrix} = \begin{vmatrix} A_i & B_i \\ C_i & D_i \end{vmatrix}, \mathrm{i} = 1, \dots, M \tag{8}$$

where $\boldsymbol{\sigma}(\boldsymbol{t})$ is a switching rule that takes values $\{\mathbf{1}, \dots, \boldsymbol{M}\}$, $\boldsymbol{M}$ is the number of subsystems. The switched control scheme is

$$u(t) = -K_{\sigma(t)}[x(t)e_I(t)]^T \tag{9}$$

To design a fuzzy switching controller, ADMIRE flight envelope has been divided into four overlapping cells as shown in **Figure 3** with the dotted lines showing the boundaries between cells. Here, the fuzzy switching control law is

$$u(t) = -K_{fuzzy\sigma(t)}[x(t)e_I(t)]^T \tag{10}$$

The controller gains are designed using the data from each related cell center, and the fuzzy switching controller is computed as follows, based on the fuzzy logic rule:

$$K_{fuzzy\sigma(t)} = \begin{cases} K_1 & Alt \leq 1550 \text{ and } Mach \leq 0.6, \\ K_2 & Alt \leq 1550 \text{ and } Mach \geq 1.1, \\ K_3 & Alt \geq 4500 \text{ and } Mach \geq 1.1, \\ K_4 & Alt \geq 4500 \text{ and } Mach \leq 0.6, \\ \rho_1 K_1 + \rho_2 K_2 & Alt \leq 1550 \text{ and } Mach \in (0.6,1.1), \\ \rho_2 K_2 + \rho_3 K_3 & Alt \in (1550,\ 4500) \text{ and } Mach \geq 1.1, \\ \rho_3 K_3 + \rho_4 K_4 & Alt \geq 4500 \text{ and } Mach \in (0.6,1.1), \\ \rho_1 K_1 + \rho_4 K_4 & Alt \in (1550,\ 4500) \text{ and } Mach \leq 0.6, \\ \rho_1 K_1 + \rho_2 K_2 + \rho_3 K_3 + \rho_4 K_4 & Alt \in (1550,\ 4500) \text{ and } Mach \in (0.6,1.1), \end{cases}$$

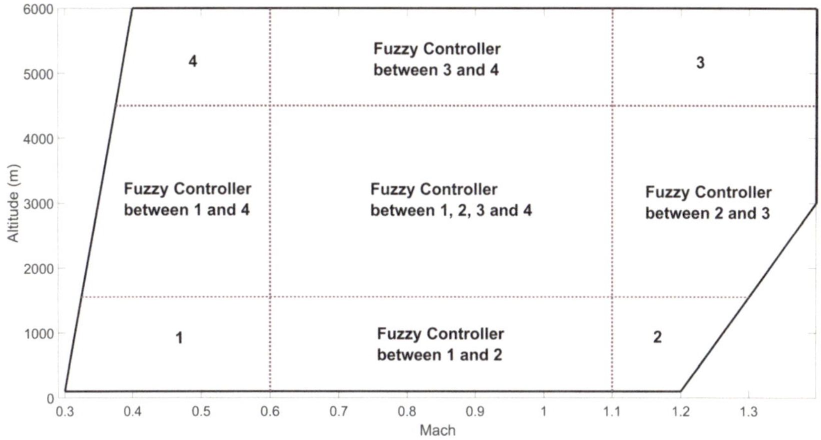

Figure 3.
Flight envelope with the four overlapping cells.

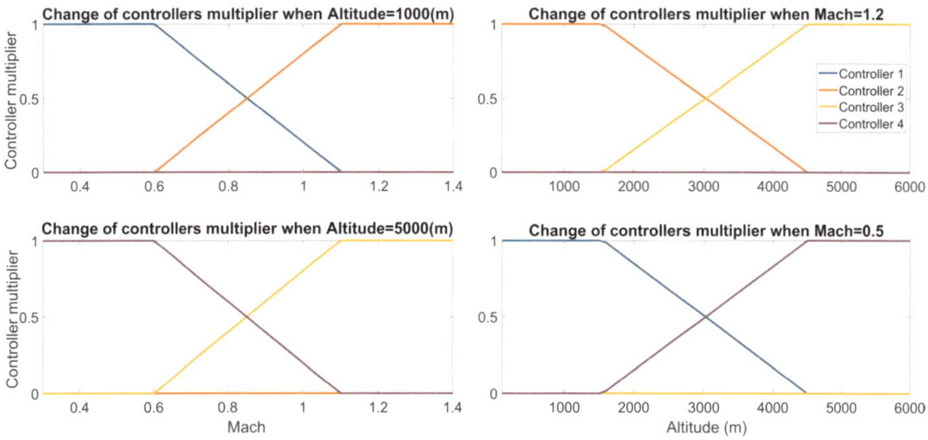

Figure 4.
Fuzzy controller rules between two cells.

where ρ_i, i [1 4] are multipliers for the related controllers as given in **Figures 4** and 5. **Figure 4** illustrates fuzzy controller rules between two cells. One can see that multiplier of the controller change linearly between active two cells, also multipliers of passive cells remain zero. In addition, the change of the controller multipliers for overlapping four cells is given in **Figure** 5.

4. Simulation results and discussion

This section represents simulation results and evaluates the performance of the developed control law using MATLAB/Simulink. Three controller strategies are compared in this section, which are the single LQI controller, the switched LQI controller given in Eq. (9), and the fuzzy switching LQI controller given in Eq. (10). The single LQI controller is designed for the data, which is taken at the center of the ADMIRE flight envelope, whereas the switched controller is designed using the flight envelope,

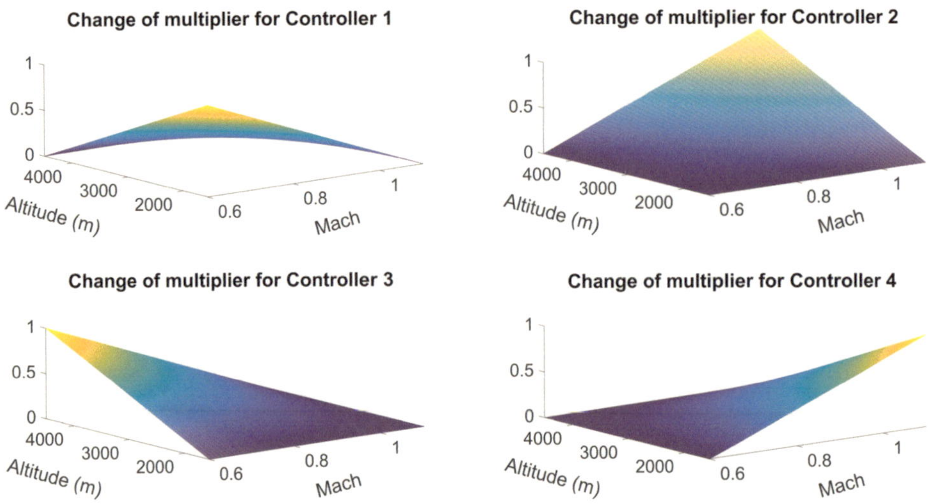

Figure 5.
Fuzzy controller rules for overlapping four cells.

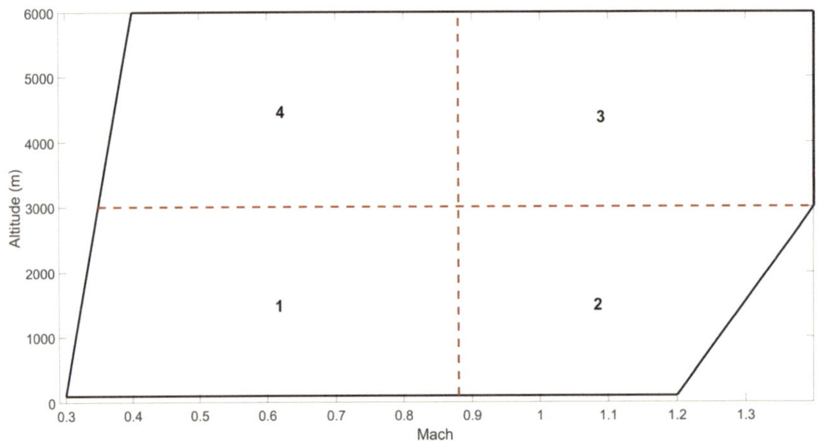

Figure 6.
Flight envelope with the four cells.

which has been divided into four cells as shown in **Figure 6** with the dashed lines showing the boundaries between cells. The feedback gains of the switched controller are computed based on the data of each cell center.

The simulation scenarios were performed to analyze the robust stability and performance of the closed-loop system at flight conditions (Mach = {0.75, 0.9, 0.75} and altitude = {5000, 1500, 4500} m). The pilot command is constricted such that the load-factor N_z stays within the design limits -3 g < N_z <9 g over the flight envelope. Load-factor demand and responses of the closed-loop system with the controllers at flight condition of Mach = 0.75 and Alt = 5000 is illustrated in **Figure 7**. The load-factor response with the single LQI is slower than the switched and fuzzy switching controllers. The switched controller has an oscillatory response during the switching, which is an undesired effect during flight operation. **Figure 8** gives the angle of attack and the Euler pitch rate responses of the closed-loop system with the single LQI, switched, and

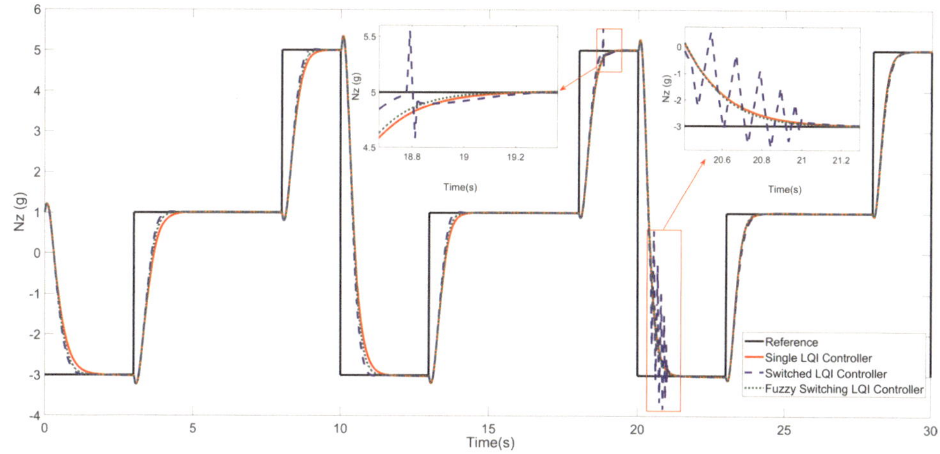

Figure 7.
Load-factor responses of the closed-loop systems at flight condition Mach = 0.75 and alt = 5000.

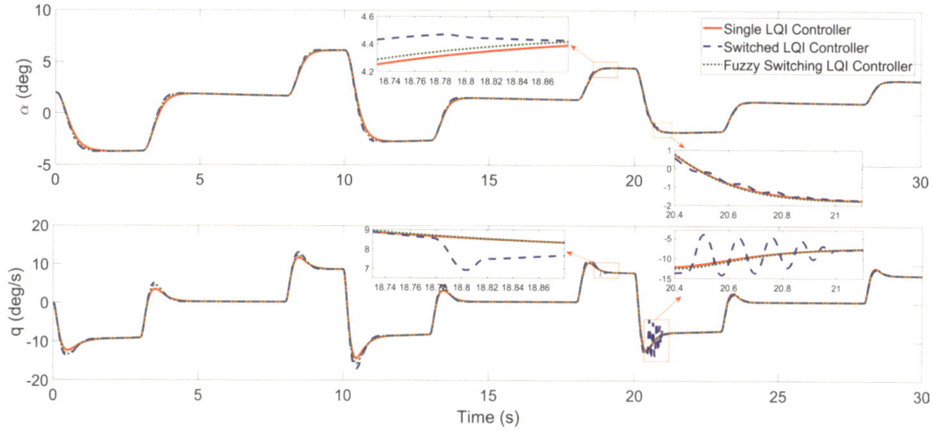

Figure 8.
State variables, the angle of attack, and the Euler pitch rate responses at flight condition Mach = 0.75 and alt = 5000.

the proposed fuzzy switching LQI controllers. One can see from the bottom plot of **Figure 8** that the Euler pitch rate response has an oscillation during switching with the switched controller. However, the proposed fuzzy switching controller has the best transient performance. The corresponding control inputs to the related controllers are given in **Figure 9**. Elevon deflection generates values between −4 deg. and 2 deg. Oscillations are also seen in this elevon deflection and throttle setting, t_{ss} when the switched controller is used. **Figures 10** and **11** give the indexes of the switched controller and the change in the coefficients of the fuzzy switching controller, respectively. Feedback gains K_1, K_2 and K_4 are employed for the switched controller, but all computed controller gains are used with the fuzzy switching controller. **Figure 12** illustrates the trajectory movement in the flight envelope for the different controllers.

In the second scenario, simulation is started at flight condition Mach = 0.9 and Altitude = 1500 m. Load-factor demand and responses of the closed-loop systems are

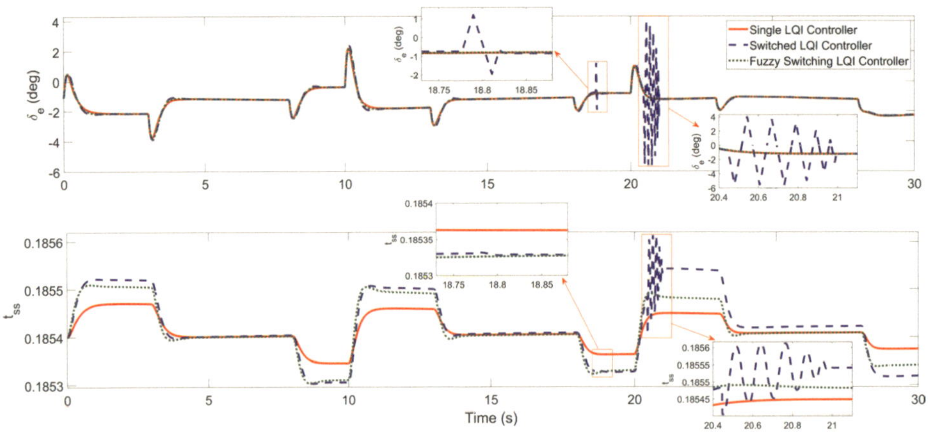

Figure 9.
Control inputs of the single, switched, and fuzzy switching controllers at flight condition Mach = 0.75 and alt = 5000.

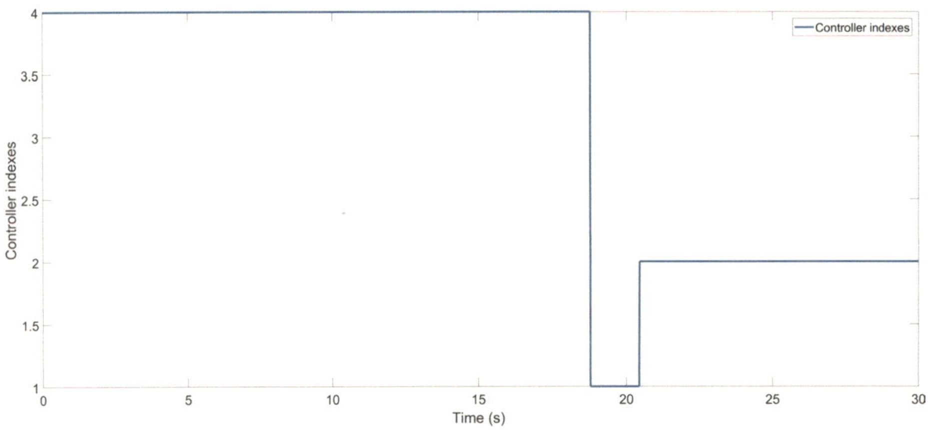

Figure 10.
Index of the switched controller gains at flight condition Mach = 0.75 and alt = 5000.

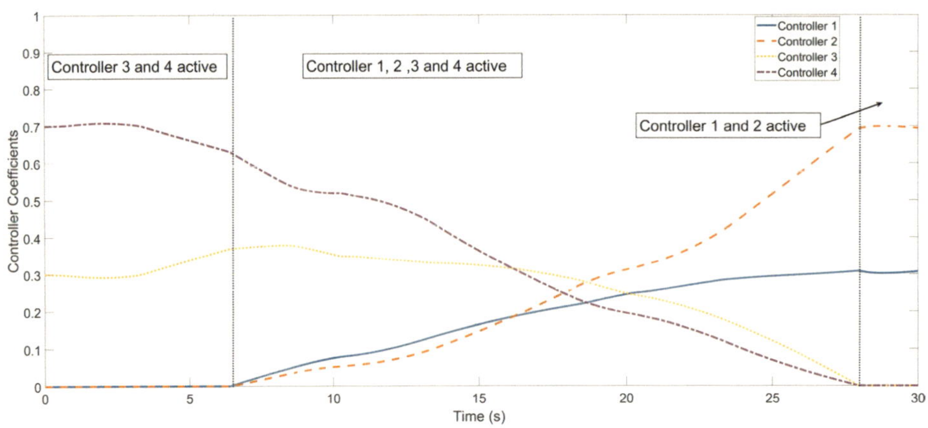

Figure 11.
Varying coefficients of the fuzzy switching controller at flight condition Mach = 0.75 and alt = 5000.

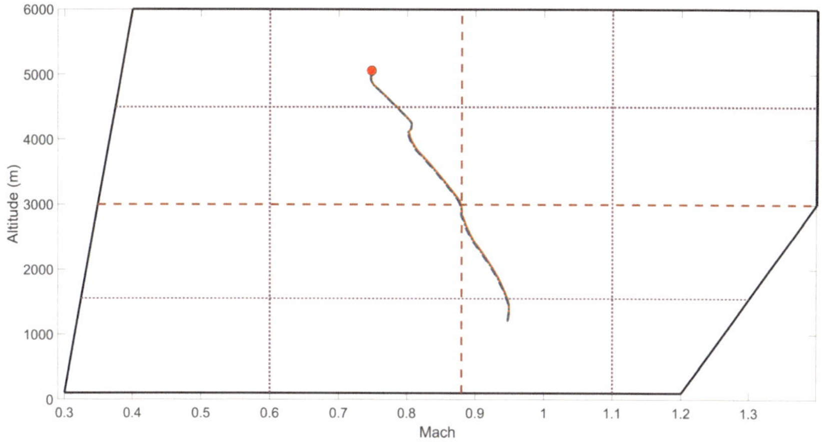

Figure 12.
Altitude responses with the different controllers at flight condition Mach = 0.75 and alt = 5000.

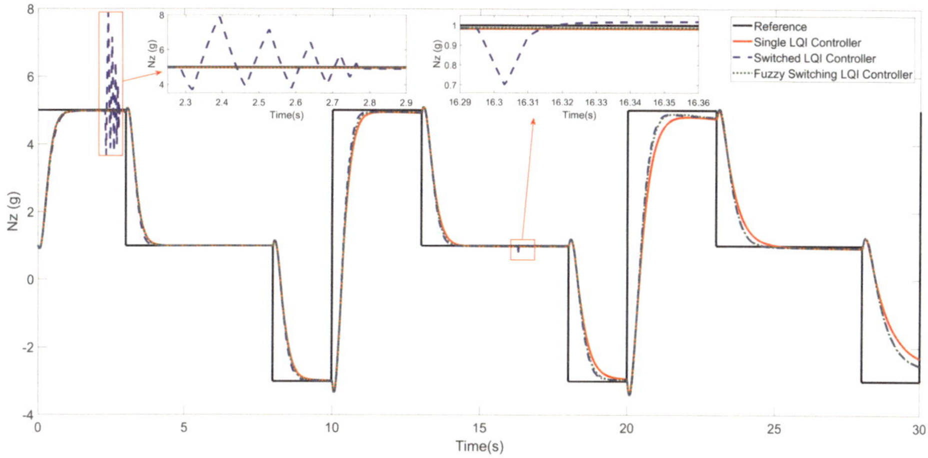

Figure 13.
Load-factor responses of the closed-loop systems at flight condition Mach = 0.9 and altitude = 1500 m.

given in **Figure 13**. Closed-loop response with the single LQI controller is the slowest amongst the controllers. Load-factor tracking response settles a larger steady-state error than the responses of other controllers. The switched controller has an undesired oscillatory response during the switching instances. **Figure 14** gives the angle of attack and the Euler pitch rate responses of the closed-loop system with the single LQI, switched, and the proposed fuzzy switching LQI controllers. The angle of attack increases at t = 20 sec for a larger demand of load-factor. Input responses of the related controllers are given in **Figure 15**. Throttle setting control input is the largest with the switched controller. The single LQI controller requires 0.288 of the throttle setting in the second scenario. **Figures 16** and **17** display the index of the switched controller and the varying coefficients of the fuzzy switching controller, respectively. All computed feedback gains are employed with the fuzzy switching controller, whereas feedback gains K_1, K_2, and K_4 are used for the switched controller. **Figure 18** illustrates the trajectory movement in the flight envelope for the different controllers.

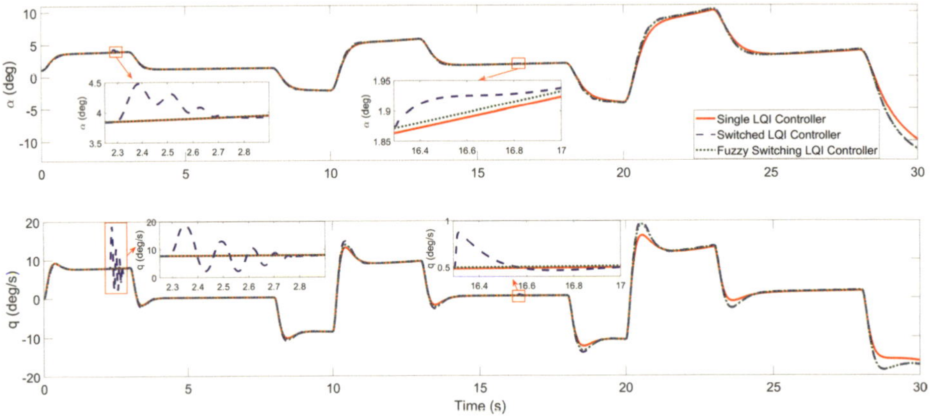

Figure 14.
State variables, the angle of attack, and the Euler pitch rate responses at flight condition Mach = 0.9 and altitude = 1500 m.

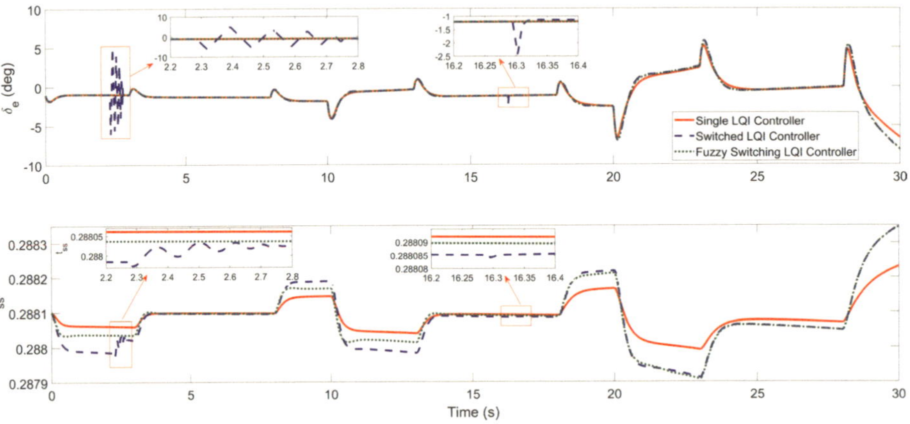

Figure 15.
Control inputs of the single LQI, switched and fuzzy switching controllers at flight condition Mach = 0.9 and altitude = 1500 m.

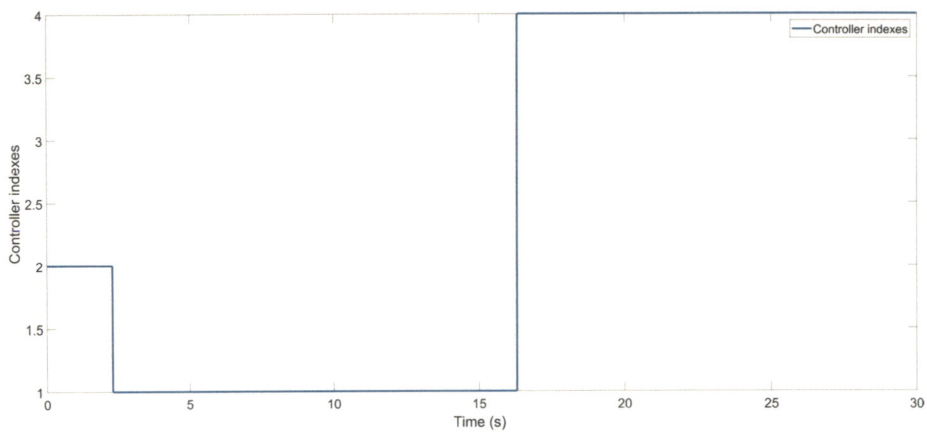

Figure 16.
Index of the switched controller gains at flight condition Mach = 0.9 and altitude = 1500 m.

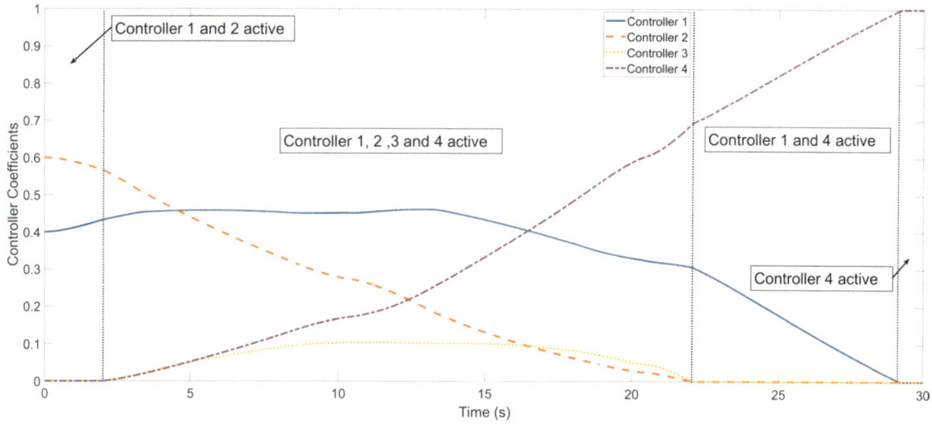

Figure 17.
Varying coefficients of the fuzzy switching controller at flight condition Mach = 0.9 and altitude = 1500 m.

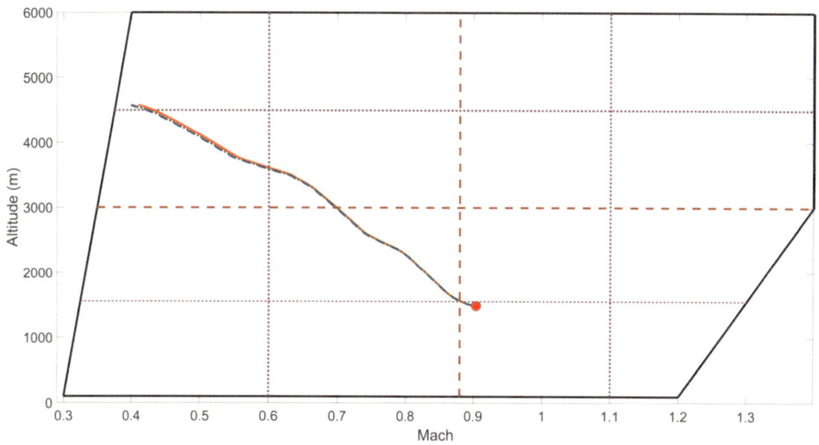

Figure 18.
Altitude responses with the different controllers at flight condition Mach = 0.9 and altitude = 1500 m.

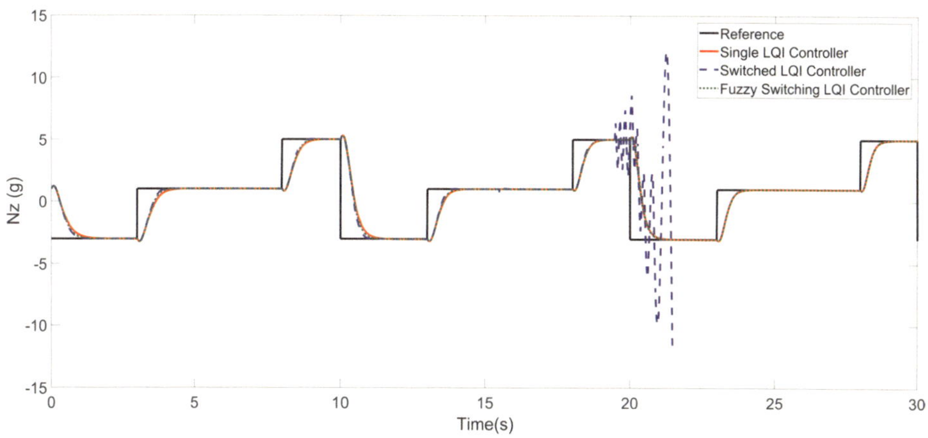

Figure 19.
Load-factor responses of the closed-loop systems at flight condition Mach = 0.75 and altitude = 4500 m.

These simulation results also demonstrate the efficacy of the proposed fuzzy switching controller.

In the third scenario, simulation is started at flight condition Mach = 0.75 and Altitude = 4500 m. Load-factor demand and responses of the closed-loop systems are given in **Figure 19**. It is clearly seen that the switched controller is unable to stabilize the aircraft when the controller switches. Load-factor tracking performance is successful with the fuzzy switching controller. The switched controller drives the closed-loop system from stability to instability shown in **Figures 20** and **21**. The index of the switched controller and the varying coefficients of the fuzzy switching controller are given in **Figures 22** and **23**, respectively. **Figure 24** illustrates the trajectory movement in the flight envelope for the different controllers. The fuzzy switching controller improves the load-factor tracking performance and enhances the stability of the aircraft.

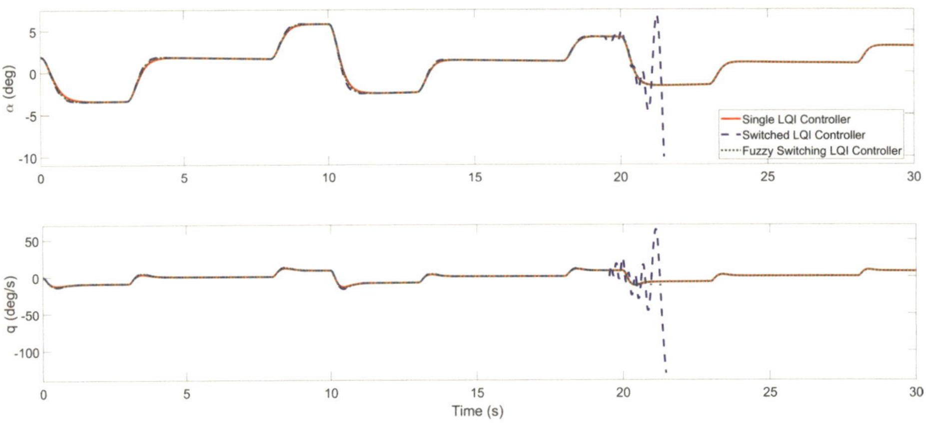

Figure 20.
State variables, the angle of attack, and the Euler pitch rate responses at flight condition Mach = 0.75 and altitude = 4500 m.

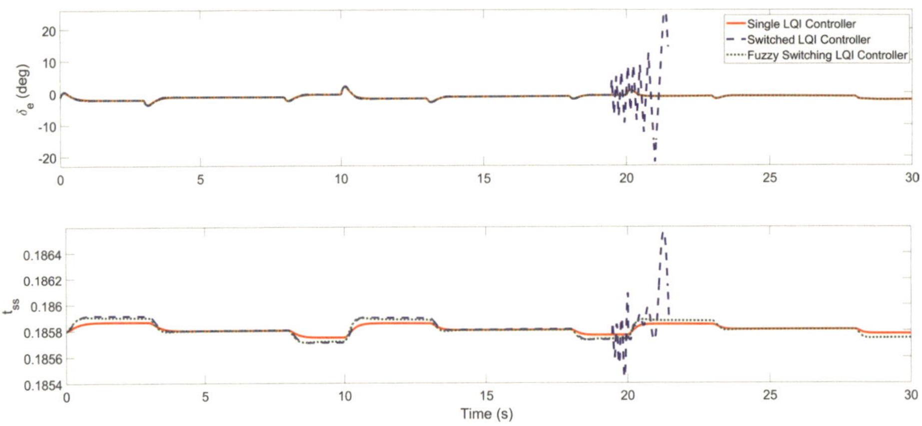

Figure 21.
Input responses of the single, switched, and fuzzy switching controllers at flight condition Mach = 0.75 and altitude = 4500 m.

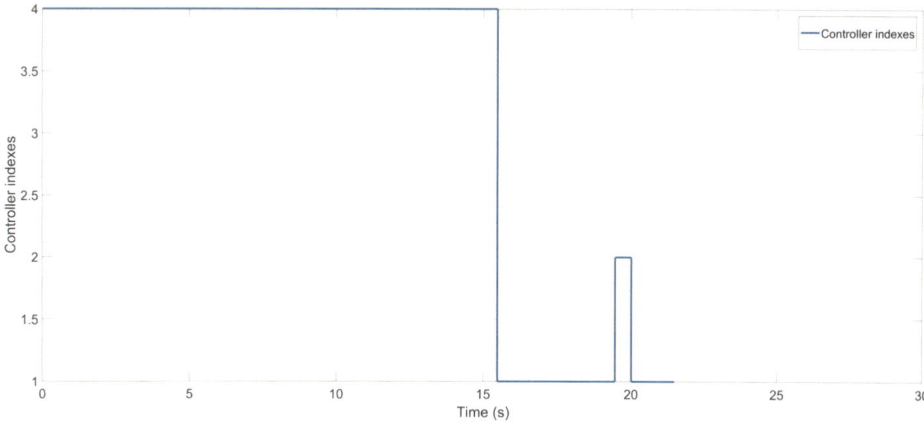

Figure 22.
Index of the switched controller gains at flight condition Mach = 0.75 and altitude = 4500 m.

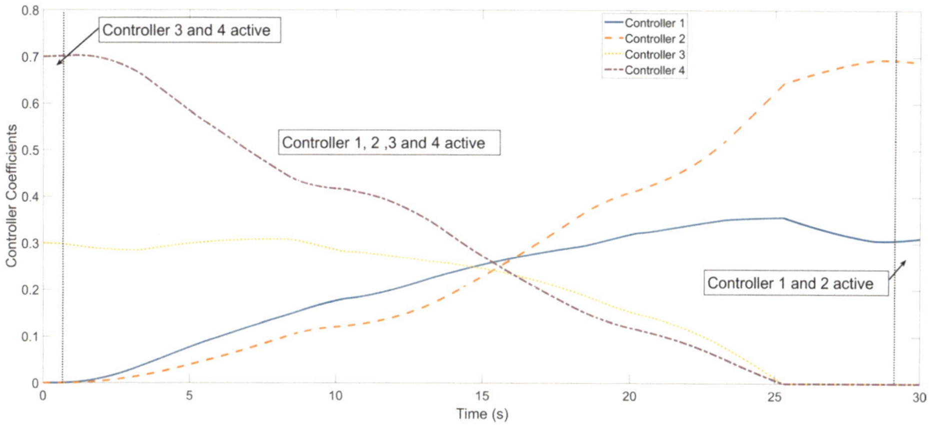

Figure 23.
Varying coefficients of the fuzzy switching controller at flight condition Mach = 0.75 and altitude = 4500 m.

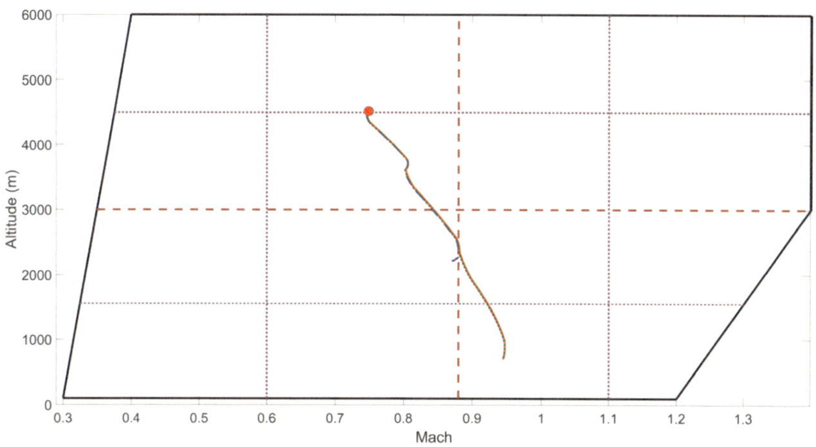

Figure 24.
Altitude responses with the different controllers at flight condition Mach = 0.75 and altitude = 4500 m.

5. Conclusions

In this chapter, a fuzzy switching controller for the ADMIRE aircraft model has been developed and verification of the control scheme was conducted using MATLAB/Simulink. Here, a switching controller is designed for the stabilization of high-performance aircraft. To improve the switching controller performance, the fuzzy logic rule has been defined and used to obtain a robust stabilization control structure instead of a single conventional LQI and the switched LQI controller.

The proposed controller scheme was compared with the standard switched and the single conventional LQI controller for load-factor tracking and robust stability under the load-factor variations. The main conclusions of the simulation results are given as follows:

- The proposed fuzzy switching controller provides better transient performance rather than the single conventional LQI and the switched controllers.
- The standard switch controller drives the ADMIRE aircraft nonlinear model from stability to instability due to switching between controllers.
- The proposed fuzzy switching controller has significant potential to improve tracking performance.
- The proposed fuzzy switching controller is effective in increasing the stability of the nonlinear system.
- Therefore, the proposed fuzzy switching controller can be preferred to control complicated and nonlinear aircraft systems. Future work will involve the stability analysis of closed-loop systems under the fuzzy switching rule.

Conflict of interest

The authors declare no conflict of interest.

Author details

Emre Kemer[1]*, Hasan Başak[2] and Hayri Baytan Özmen[3]

1 Department of Electrical and Electronics Engineering, Uşak University, Uşak, Turkey

2 Department of Electrical and Electronics Engineering, Artvin Çoruh University, Artvin, Turkey

3 Department of Civil Engineering, Uşak University, Uşak, Turkey

*Address all correspondence to: emre.kemer@usak.edu.tr

References

[1] Zhang Y, Suresh VS, Jiang B, Theilliol D. Reconfigurable control allocation against aircraft control effector failures. In: Proceedings of the IEEE International Conference on Control Applications; 1–3 October 2007. Singapore: IEEE; 2007. pp. 1197-1202

[2] Han Y, Li P, Ma J. Vector-coupled flight controller design based on multivariable backstepping sliding mode. Symmetry. 2019;**11**(10):1225

[3] Zhang Q, Sijun Y, Yan L, Xinmin W. An enhanced LMI approach for mixed H2/ H∞ flight tracking control. Chinese Journal of Aeronautics. 2011;**24**(3):324-328

[4] Ochi Y. Design of a flight controller for hypersonic flight experiment vehicle. Asian Journal of Control. 2004;**6**(3): 353-361

[5] Zhang L, Bi S, Yang H. Fuzzy-PID control algorithm of the helicopter model flight attitude control. In: Proceedings of the IEEE Chinese Control and Decision Conference; 26-28 May 2010. Xuzhou, China: IEEE; 2010. pp. 1438-1443

[6] Herrmann AA, Ben-Asher JZ. Flight control law clearance using optimal control theory. Journal of Aircraft. 2016; **53**(2):515-529

[7] Bouvier J-B, Ornik M. Designing resilient linear systems. IEEE Transactions on Automatic Control. 2022;**67**(9):4832-4837

[8] Da Costa RR, Chu QP, Mulder JA. Reentry flight controller design using nonlinear dynamic inversion. Journal of Spacecraft and Rockets. 2003;**40**(1):64-71

[9] De Almeida FA. Robust off-line control allocation. Aerospace Science and Technology. 2016;**52**:1-9

[10] Sonneveldt L, Chu QP, Mulder JA. Nonlinear flight control design using constrained adaptive backstepping. Journal of Guidance, Control, and Dynamics. 2007;**30**(2):322-336

[11] An H, Fidan B, Wu Q, Wang C, Cao X. Sliding mode differentiator based tracking control of uncertain nonlinear systems with application to hypersonic flight. Asian Journal of Control. 2019; **21**(1):143-155

[12] Li P, Ma J, Zheng Z. Disturbance-observer-based fixed-time second-order sliding mode control of an air-breathing hypersonic vehicle with actuator faults. Proceedings of the Institution of Mechanical Engineers, Part G: Journal of Aerospace Engineering. 2018;**232**(2): 344-361

[13] Ma J, Li P, Zheng Z. Disturbance observer based dynamic surface flight control for an uncertain aircraft. Proceedings of the Institution of Mechanical Engineers, Part G: Journal of Aerospace Engineering. 2018;**232**(4): 729-744

[14] Su X, Chen CP, Liu Z. Adaptive fuzzy control for uncertain nonlinear systems subject to full state constraints and actuator faults. Information Sciences. 2021;**581**:553-566

[15] Abbasi SMM, Jalali A. Fuzzy tracking control of fuzzy linear dynamical systems. ISA Transactions. 2020;**97**: 102-115

[16] Luan T, Sun M, Hu Z, Fu Q, Wang H. A novel TS fuzzy robust control for part transportation of aircraft carrier considering transportation time and stochastic demand. Aerospace Science and Technology. 2021;**119**: 107096

[17] Yahui W, Yitao L, Yuhuan L, Ying N, Li L, Zhengrong L, et al. Adaptive fuzzy control for input restriction airbreathing hypersonic vehicle. Journal of Physics: Conference Series. 2021;**1738**:012084

[18] Forssell L, Nilsson U. ADMIRE the aero-data model in a research environment version 4.0, model description. Technical report FOI-R-1624-SE FOI. 2005

[19] Sidoryuk ME, Goman MG, Kendrick S, Walker DJ, Perfect P. An LPV control law design and evaluation for the ADMIRE model. In: Nonlinear Analysis and Synthesis Techniques for Aircraft Control. Berlin, Heidelberg: Springer; 2007. pp. 197-229

[20] Simon D. Fighter Aircraft Maneuver Limiting Using MPC: Theory and Application. Linköping: Linköping University Electronic Press; 2017. p. 17

Chapter 9

Adaptive Neuro-Fuzzy Inference System-Based GPS-IMU Data Correction for Capacitive Resistivity Underground Imaging with Towed Vehicle System

Elmer Dadios, Jonah Jahara Baun, Mike Louie Enriquez, Adrian Genevie Janairo, Ronnie Concepcion II, Joseph Aristotle De Leon, Kate Francisco, Andres Philip Mayol, Argel Bandala and Ryan Rhay Vicerra

Abstract

This study proposes the utilization of an Adaptive Neuro-Fuzzy Inference System (ANFIS) to correct the latitude and longitude of Global Positioning System (GPS) used in locating towed vehicle system for underground imaging. The input used was the collected data from a developed Real-time Kinematic Global Positioning System sensor integrated with Inertial Measurement Unit. Different ANFIS models were developed and evaluated. For latitude correction, ANFIS model with hybrid optimization trained at 300 epochs was chosen, whereas for longitude correction, ANFIS model with hybrid optimization trained at 100 epochs was selected. Both models achieved the lowest Mean Squared Error (MSE), the highest Coefficient of Determination (R^2), and lowest Mean Absolute Error (MAE). Moreover, selected best ANFIS models were compared to Long Short-Term Memory (LSTM) and Extreme Learning Machine (ELM) models, but the results showed that the ANFIS models have superior performances. The selected ANFIS models were verified by testing on the collected actual dataset and the visualized map demonstrated that the corrected GPS latitude and longitude have significantly reduced error, indicating that the fuzzy system with neural network capabilities is a cost-effective and convenient method for error reduction in vehicle localization making it applicable to be integrated for capacitive resistivity underground imaging systems.

Keywords: adaptive neuro-fuzzy inference, global positioning system, inertial measuring unit, underground imaging, towed vehicle

1. Introduction

Underground imaging is a noninvasive approach for creating a subsurface area model by transmitting low current, high alternating voltage, and low-frequency waves into the ground. This method has various applications such as detecting minerals, finding underground materials and faults, and detecting voids [1]. One of the techniques used in underground imaging is the capacitive resistivity (CR) method [1–3], where the ground-coupled transmitter and receiver antennas are designed and configured in a capacitive connection to measure resistivity by determining the potential difference [4–6]. The design of a capacitive resistivity imaging system can be either a single-pair antenna system that includes a transmitter and a receiver unit, or a multiple-antenna system with one transmitter and multiple receiver units [7]. Both systems use a vehicle towing mechanism during the surveying process, as presented in **Figure 1**, allowing for the mapping and location of buried utilities [8].

Accordingly, the CR method employs map-matching (MM) algorithms that utilize positioning sensors, such as the Global Positioning System (GPS) in combination with digital maps to determine the road link on which a vehicle is traveling and to obtain highly accurate data for mapping. GPS-based data collection is more efficient than traditional surveying and mapping approaches, requiring less equipment and labor [9]. It offers direct information on latitude and longitude coordinates without the need for angle and distance measurements between intermediary points. Despite its widespread use, GPS has some limitations that should be considered. For instance, GPS units cannot always provide locations with high accuracy beyond 3 meters, which could be problematic for certain applications. In addition, GPS devices rely on signals from at least four satellites to precisely pinpoint a location, and signal blocking or interference, such as in urban areas or under tree canopies, can impact performance. GPS accuracy may also be affected in environments with high ionospheric activity, such as during solar flares or geomagnetic storms. Finally, raw, uncorrected GPS data

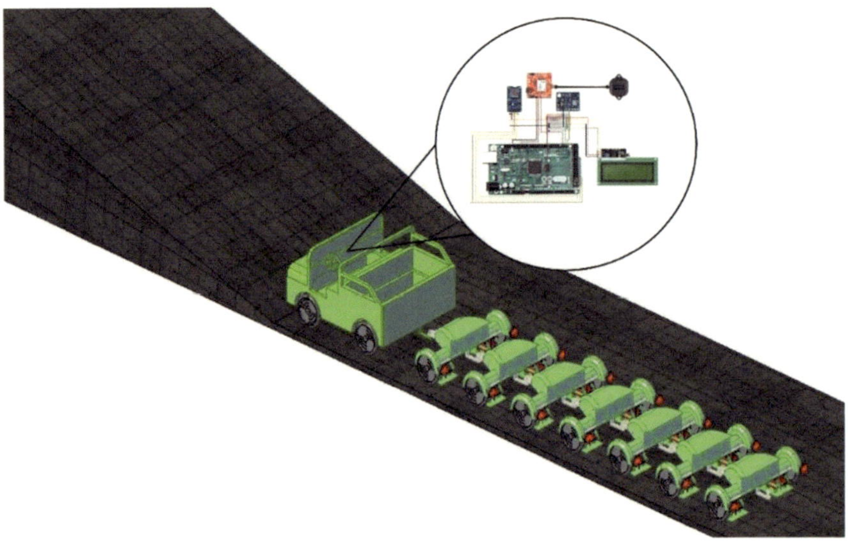

Figure 1.
The placement of the GPS-IMU device utilized for localization and tracking of the capacitive resistivity towed antenna system used for underground imaging is illustrated.

points may only be precise up to 5 meters, and a clear view of the sky is necessary to receive signals from GPS satellites [10, 11]. In capacitive resistivity imaging, lowering GPS accuracy error is critical for finding underground utilities and performing map matching, however, the mapping technology used in conjunction with GPS may not always be up-to-date or reliable, potentially leading to navigational errors [12, 13].

One known method to overcome the inaccuracy of mapping GPS sensor data alone is the IMU-GPS sensor fusion. This technique combines data from a GPS receiver and an Inertial Measurement Unit (IMU) to improve the accuracy and robustness of navigation and positioning systems [14]. It provides data regarding the orientation and acceleration of the device, while GPS provides information about its absolute position. By integrating the data from both sensors, the position and orientation estimates are more precise and reliable than the readings obtained from each sensor separately [14, 15].

In contrast to the state-of-the-art, which typically employs more conventional methods for localizing and land vehicle tracking, for instance, the Kalman filter, fuzzy logic is considered a commonly known artificial intelligence (AI) approach. Researchers in [16] created a strong Kalman filter utilizing vector tracking and integrated it with fuzzy logic to change filter parameters to follow weak signals in global navigation satellite system (GNSS) receivers. Thus, the results were superior to the standard procedure. Following this work, a fuzzy position correction method for latitude and longitude data from a GPS sensor was introduced, which was implemented on Field-Programmable Gate Arrays (FPGA) to speed up rectification results. Compared to other models, the FPGA-based approach provided a 40,000× speedup [17]. Combining antenna optimization techniques and sensor fusion with AI has been introduced to increase GPS accuracy [18]. Even when employing an inexpensive GPS sensor for location-based applications, this effectively computed correct latitude and longitude data. In another study, the authors utilized an unscented Kalman filter (UKF) and an unscented H-infinity (UH) filter to track ground vehicle position using fuzzy logic to decide which to use at any given time, lowering error by 5.6% and enhancing GPS accuracy [19]. Moreover, a fuzzy system model was developed [20] that flexibly adjusts the noisy covariance values of the extended Kalman filter (EKF) by combining data from GPS, an odometer, IMU, and the automobile's mathematical framework. This results in a 49% improvement in the precision of the vehicle's absolute position [20]. Similarly, Zhu et al. employed EKF to fuse data from a four-wheeled robot's GPS, IMU, odometers, and camera. They created a fuzzy system to adjust the noisy covariances of the EKF. The strategy successfully improves the robot's estimation of the trajectory to be followed by 80.6% [21]. These studies suggest that fuzzy logic has a great potential to be utilized for land vehicle tracking and localization, providing higher accuracy than the traditional approaches.

One type of artificial intelligence that integrates the capabilities and strengths of both neural networks and fuzzy logic systems is the adaptive neuro-fuzzy inference system (ANFIS), which is used for analyzing input–output relations, can handle inaccurate or imperfect data and has been useful to various domains such as in pattern recognition and predictive modeling [22]. Specifically, ANFIS has also been employed in the localization and tracking of land vehicles and shows significant findings in different scenarios. The work in [23] classified direct, multipath-affected, and non-line of sight (NLOS) findings using raw GPS measurements with an ANFIS algorithm. The correct classification rates were 100, 91, and 84%, respectively. Another study proposed an intelligent ANFIS system that modifies the position of a vehicle based on sensor data and latitude/longitude. According to the results, the fuzzy system outperformed the unscented Kalman filter by 69.2% [24]. Another study employs ANFIS to estimate an

IMU's inaccuracy over time. While it refrains from specifically addressing GPS-IMU incorporation, it offers details on the application of ANFIS for IMU data error estimation, and the outcome implies that the ANFIS could substantially enhance the accuracy of inertial navigation positioning, which is important for vehicle inertial navigation in intricate and covert settings [25]. Research by [26] presents an ANFIS-based approach to categorizing everyday life events using data collected by IMU sensors. Although it concentrates on identifying activities rather than GPS-IMU data correction, it still exhibits the usage of ANFIS in sensor fusion with a total accuracy of 98.88%. The real-time deployment of ANFIS for vehicle navigation is proposed in [27]. When evaluated on a vehicle, the suggested model outperforms standard methods in terms of accuracy. The experimental findings proved the benefits of the suggested AI-based INS/GPS integration strategies in terms of robustness while maintaining real-time system location accuracy [27]. Although ANFIS has several advantages for GPS-IMU localization and vehicle tracking, it is vital to evaluate the potential limits and challenges of employing this technology. In [24, 28], the study showed that one potential limitation of ANFIS is that it may necessitate a huge amount of data to accurately train the model, and ANFIS models can be complex and hard to understand, which renders it more challenging to comprehend how the model works and makes decisions, particularly for autonomous vehicle applications. Thus, future research still requires more exploration of ANFIS modeling to prove its advantages and disadvantages in certain fields.

In relation to this proposed study, the main objective is to correct the GPS sensor's latitude and longitude coordinates to avoid complex mathematical operations and achieve a comprehensive location system embedded in a towed antenna system. Thus, it is critical to reduce the error in the accuracy of GPS receivers, which ensures the correct location of underground utilities and for map matching purposes. With that, this study aims to propose an intelligent system-based fuzzy logic using ANFIS, which takes the information from Real-time Kinematics (RTK) GPS sensors with integrated IMU's linear acceleration and orientation data, which corrects the capacitive resistivity imaging system vehicle's absolute position according to its latitude and longitude. This correction uses two fuzzy systems, one for latitude and the other for longitude, which will be trained using the ANFIS tool. The positioning correction system will be trained and tested with datasets from constructed Arduino-based RTK-GPS with Integrated IMU. Moreover, the developed models are compared to the performance of two neural network models – long short-term memory (LSTM) and extreme learning machine (ELM) for the comparison and validation of the proposed method. This research is expected to provide significant benefits, including (1) facilitating the integration of different measurement modalities and improving the interpretation and visualization of the data [29] which can enhance the understanding of the subsurface properties being investigated in capacitive resistivity imaging (CRI) systems, (2) aid in mapping the precise location of the underground utility objects being surveyed by the underground imaging towed antenna system, and (3) the ability to automate data acquisition and processing, which can save time and reduce errors and provide more precise control over the measurement process, enabling more accurate and reproducible results.

2. Materials and methods

The proposed step-by-step procedures of how to correct the collected GPS sensor's latitude and longitude coordinates used to find the position of the towed array vehicle in an underground imaging system are presented in **Figure 2**. This provides a

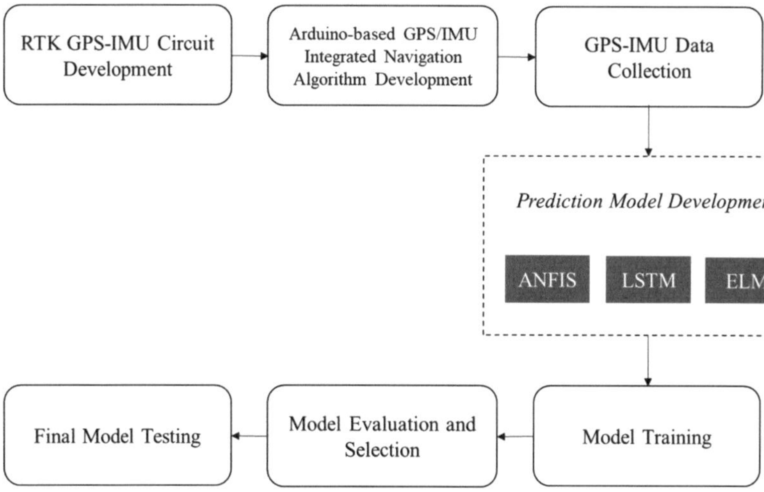

Figure 2.
Step-by-step process of GPS sensor's latitude and longitude coordinates correction used for the capacitive resistivity underground imaging system.

functional view of how the whole research system works. The process starts with the hardware circuit development of an RTK-GPS sensor with an integrated IMU sensor, followed by the incorporation of a developed Arduino-based GPS/IMU integrated navigation algorithm to complete the device needed for GPS and IMU data collection, which is conducted through actual field testing. After collecting the data, three prediction models were developed: ANFIS, LSTM, and ELM, while the collected data were trained using these three models to predict the corrected GPS latitude and longitude. The prediction model with the highest accuracy was selected while LSTM and ELM prediction models were also used to validate the performance of ANFIS and, thus, utilized for final model testing of collected data.

2.1 Arduino-based RTK-GPS with integrated IMU

An Arduino-based RTK-GPS with an integrated IMU is a device that combines a real-time kinematic (RTK) global positioning system (GPS) with an inertial measurement unit (IMU) using an Arduino microcontroller. The overall block diagram of how the RTK-GPS is combined with the IMU sensor is presented (**Figure 3**) to comprehend the overall system architecture and information flow between various functional parts. The RTK-GPS utilized is a Sparkfun SMA-ZED-F9P model, which is the highest-quality module for high-accuracy GNSS and GPS navigation solutions, including RTK, with 10 mm three-dimensional accuracy, while the utilized IMU sensor is GY-85 9DOF Sensor that has nine axes which are triaxial gyroscope, triaxial accelerometer, and triaxial magnetic field. The RTK-GPS sensor can calculate the satellite position and velocity and provides a GPS raw pseudo-range that is the pseudo-distance between the satellite and GPS receiver, and the pseudo-rate specifies the velocity. Through this, the RTK-GPS sensor strengthens GPS signals for exact locations and velocities. On the other hand, the IMU sensor performs the coordination rotation, which is the process of aligning the axes of the IMU sensor with the axes of the vehicle where it is mounted. It also performs altitude determination. These two are significant aspects of sensor fusion in localization and vehicle tracking. Then, by

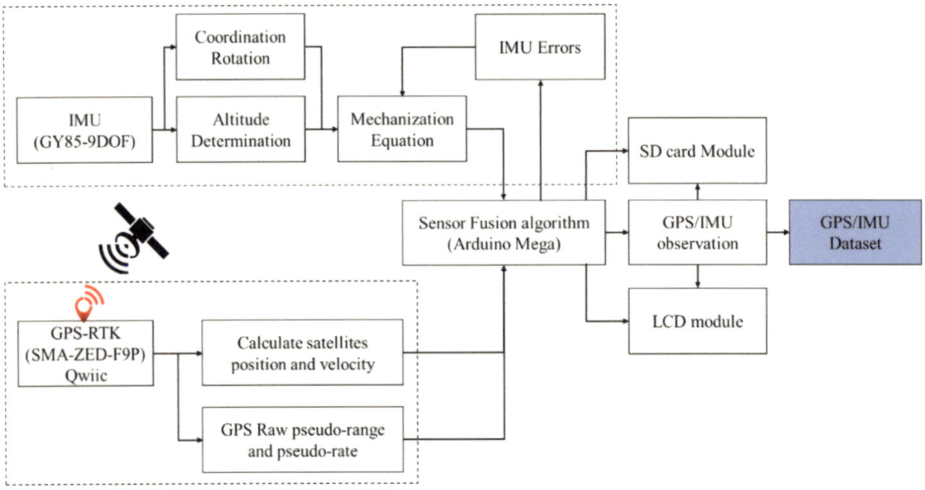

Figure 3.
Block diagram of the development of Arduino-based GPS-RTK with integrated IMU for sensor fusion.

utilizing mechanization equations, the accelerometer, gyroscope, and magnetometer in the IMU sensor module offer information on linear acceleration, angular velocity, and magnetic field strength along three axes, also considering the estimation errors present in the computation process. If IMU errors are not dealt with or assessed correctly as part of a combined GPS-IMU system, these can cause major inaccuracies in location, velocity, and attitude calculation. To build the system, the RTK-GPS sensor, IMU sensor, Secure Digital (SD) card module, and LCD monitor are connected to the Arduino mega board using various interfaces such as serial peripheral interface (SPI), inter-integrated circuit (I2C), and universal asynchronous receiver/transmitter (UART). Once the components are connected, necessary libraries [30] are installed, and the algorithm for sensor fusion is written to read data from the RTK-GPS and IMU, store the data on the Secure Digital (SD) card, and display the data on the LCD monitor in real-time. Thus, providing an output of the GPS-IMU dataset.

To show the actual electrical connections between the components in the circuit, the electronic circuit diagram is presented (**Figure 4**) to create a powerful system that can provide high-precision positioning data and information about the device's orientation and movement.

Overall, the RTK-GPS component provides high-precision positioning data [31, 32], while the IMU [32–34] component provides information about the device's orientation and movement. Combining these two components can produce more accurate and reliable data than either component alone. By combining the functionalities of the RTK-GPS and IMU, the device provides more accurate and reliable data than either component alone [34]. As a result, it is a vital tool for a broad spectrum of applications requiring precise location and movement data. The setup function initializes the modules, sets the output rate of the RTK-GPS module to 20 Hz, creates a log file on the SD card, and initializes the 20 × 4 LCD. The loop function reads data from the RTK-GPS module, converts it into a more readable format, and writes it to the log file and LCD display. The data logged includes the latitude, longitude, mean sea level and accuracy. In addition, the code was designed to read data from the IMU sensor module, which contains an accelerometer, a magnetometer, and a gyroscope. The data includes acceleration, magnetic field strength, and angular velocity in three axes, as well as gyroscope temperature.

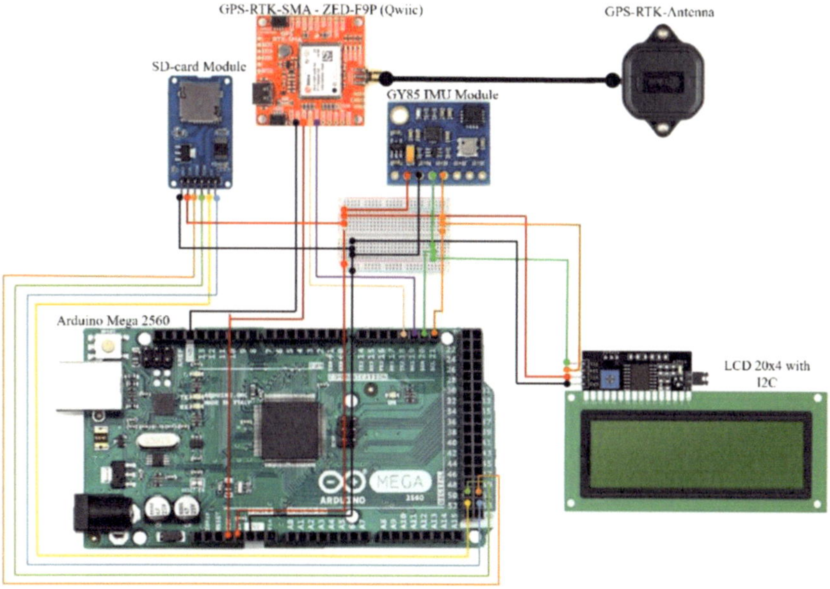

Figure 4.
Circuit diagram of Arduino-based GPS-RTK with integrated inertial measurement unit.

2.2 Integrated GPS-IMU data collection

Using the Arduino-based integrated GPS-IMU device with sensor fusion algorithm, a total of 2521 row data of GPS latitude and longitude, triaxial accelerometer, and triaxial magnetometer information were collected. The device was mounted in front of a testing vehicle while the circuit testing was conducted, starting from a specific point at Lumban, Laguna, and ending at De La Salle University (DLSU) Manila Campus.

The data from RTK-GPS with an integrated IMU sensor is acquired and logged through the connected SD card in the Arduino, and it automatically saves all the collected data in Comma Separated Values (CSV) files to be able to present the data in tabular format and efficient data processing. To verify if the collected actual data has significant errors, the collected GPS latitude and longitude data from the testing route were plotted into a map, as seen in **Figure 5** (in red pin markers). The plotting of GPS latitude and longitude coordinates started by converting the saved CSV file into GPS eXchange Format (GPX) to properly store, exchange, and map GPS location data. MyGeodata Cloud is used in the conversion; this is an online converter tool that allows users to convert CSV files to GPX format in batch. Using Google Maps, the GPX file containing the collected GPS latitude and longitude is imported, and the map way-points are automatically added and plotted. For the reference GPS points of latitude and longitude, the reference route is extracted from Google Maps and then converted to a GPX file. This GPX file is also plotted in Google Maps, as seen in **Figure 5** (the blue line), for comparison and visualization. The collected reference GPS latitude and longitude data were also utilized as the target output for the prediction models. Thus, **Figure 5** clearly represents the reference GPS data in the blue line and the plotted collected GPS data in red pin markers. From **Figure 6**, it is evident that the collected GPS data are not close to the reference route (blue line), which shows that there are inconsistencies and errors with the collected GPS data. The error is too large, which should be corrected using the ANFIS tool.

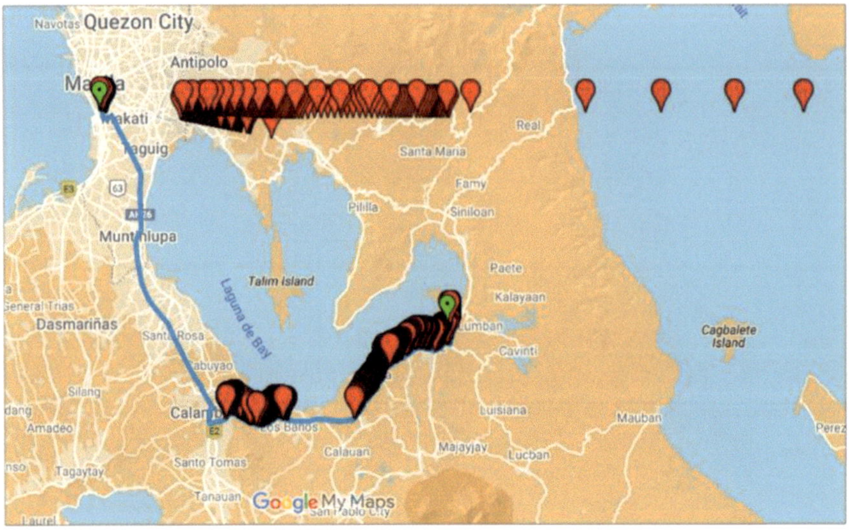

Figure 5.
Mapped collected GPS data points (red pin markers) and reference route (blue line) from Lumban, Laguna to DLSU.

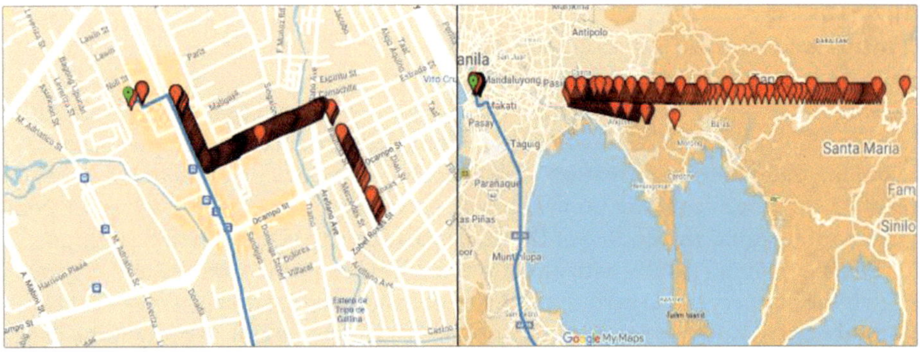

Figure 6.
Inconsistencies and errors of collected GPS data (red pin markers) with respect to the reference route (blue line).

2.3 Adaptive neuro-fuzzy inference system modeling

ANFIS, an Adaptive Neuro-Fuzzy Inference System, is an intelligent system that combines the capabilities of Artificial Neural Network (ANN) and Fuzzy Logic Inference System (FIS) to bridge the gap that exists between the two [35–37]. This is a well-established technique that employs relational models to represent linear and nonlinear relationships between input and output parameters, taking into account the fact that human knowledge is often fuzzy and not strictly defined [35]. The function of the human nervous system is depicted by the Fuzzy Inference System (FIS), which is supported by the Artificial Neural Network (ANN). A neuro-fuzzy component forms each layer of the ANFIS, which can be recognized as a feedforward ANN that was developed by [36]. The input variables' activation process will take place via the function parameters, which are trained using an optimization method defined by the input membership function (MF) and then passed on to the next neuron. Following this, the activation values will be identified by the fuzzy rules and sent to the output MF before being transferred to the output node [35].

Through ANFIS, a fuzzy system may correct several inputs at the same time, and these multiple inputs employed in the ANFIS model are GPS latitude, GPS longitude, 3-axis point coordinates of the accelerometer, and 3-axis point coordinates of the magnetometer. Two fuzzy systems are developed in this case, one for correcting the latitude and the other for longitude correction. The combined GPS and IMU sensors provided the same data to both fuzzy systems. The data used for training and testing are collected data through the GPS-RTK with an integrated IMU sensor. In particular, the subclustering method was utilized for FIS generation in the two models-latitude and longitude, as it demonstrated the best performance during algorithm pre-evaluation. This method generates a Sugeno-type FIS structure that is utilized to initialize the membership function parameters (**Figures** 7 and **8**) [38]. This method involves dividing the input space into several subsets, known as clusters, and identifying the optimal number of clusters required to represent the input space accurately [39]. Once the clusters are identified, the corresponding membership function parameters are initialized based on the cluster centers and widths. This initialization allows for faster and more accurate training of the ANFIS network using two different optimization methods.

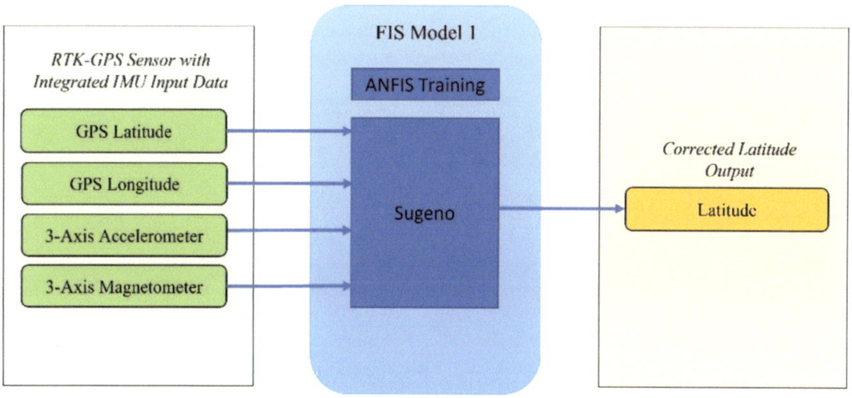

Figure 7.
Sugeno-type FIS for training the latitude ANFIS model.

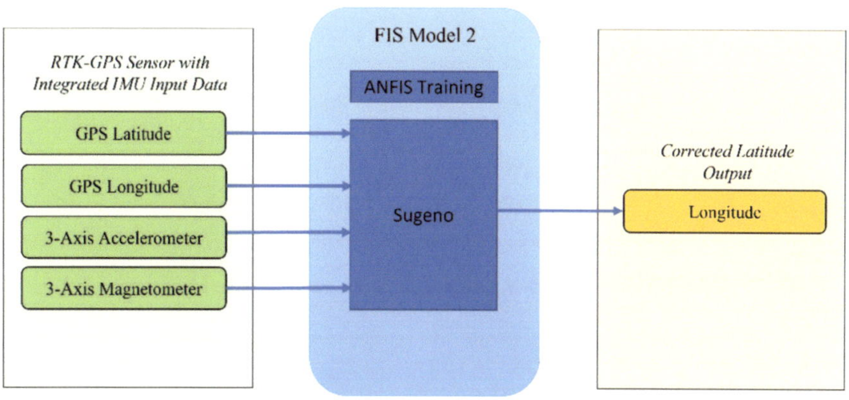

Figure 8.
Sugeno-type FIS for training the longitude ANFIS model.

In the ANFIS tool, the range of influence was configured to 0.25 to allow the model to create smaller data clusters and produce more fuzzy rules. The squash factor is 0.6 to also create more and smaller data clusters. The acceptance ratio value is 0.25, which is greater than the rejection ratio value of 0.15. For each of the two models, four models are generated by changing the number of epochs of either 100 or 300 and altering the optimization methods used-hybrid, which is the combination of back-propagation and least mean squared error method and backpropagation. The best model will be selected based on which has the lowest error in training and testing.

For the latitude correction ANFIS model (**Figure 9**), the resulting ANFIS structure has eight inputs and one output node with 21 input MF for each input, generating 21 fuzzy rules and 389 nodes. The eight inputs represent the collected GPS latitude and longitude data points, the 3-axis coordinates of the accelerometer, the 3-axis coordinates of the magnetometer, and the reference latitude as the targeted output. The rest of the other latitude models followed the same structure and configuration of the network.

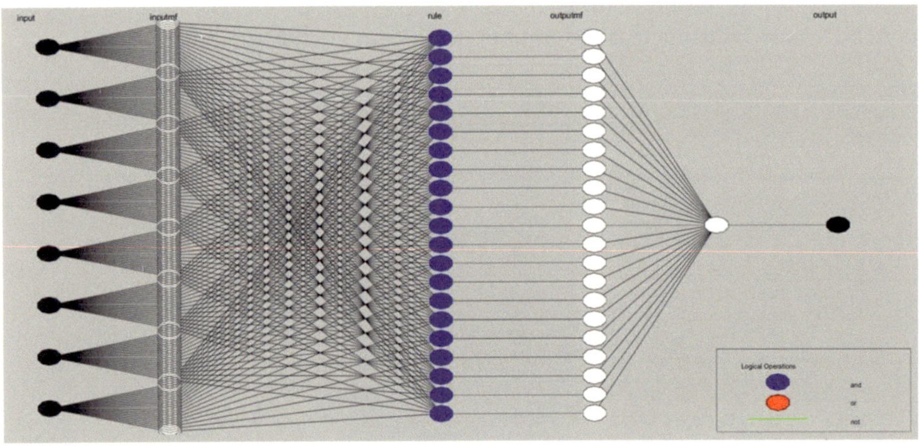

Figure 9.
ANFIS subclustering network structure for corrected latitude prediction.

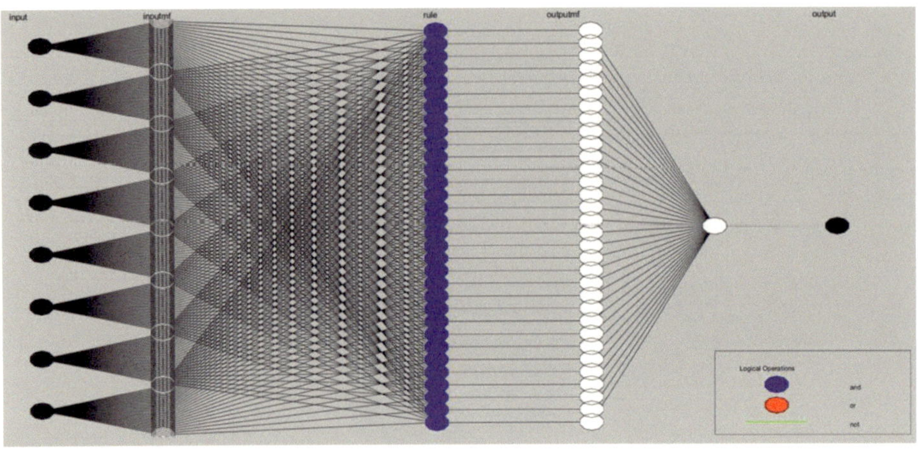

Figure 10.
ANFIS subclustering network structure for corrected longitude prediction.

Moreover, **Figure 10** shows the resulting ANFIS structure of the longitude correction ANFIS model that has eight inputs-the collected GPS latitude and longitude datapoints, 3-axis coordinates of the accelerometer, and 3-axis coordinates of the magnetometer, and reference longitude as the target output with 32 input MF for each input allowing to producing of 32 fuzzy rules and 439 nodes. All the simulated longitude models followed the same structure and configuration of the network.

2.4 Long short-term memory (LSTM) and extreme learning machine (ELM) modeling

A special kind of Recurrent Neural Network that can recognize long-term dependencies is known as Long Short-Term Memory Network or LSTM. LSTMs are designed specifically to avoid the long-term dependence problem [40]. Their behavior is set up to make retention of memory over time their default setting. Therefore, they perfectly compare time series forecasting [40]. This study employs LSTM to develop prediction models for latitude and longitude correction. The hyperparameters used in modeling are presented in **Table 1**. A three-layer LSTM network was simulated with different combinations of hidden neurons on each layer. Hidden layer 1 comprises 500, 1000, and 1500 hidden neurons; hidden layer 2 has 700, 500, and 300 hidden neurons, while hidden layer 3 shall consist of 300, 200, and 100 hidden neurons. These different combinations of hidden neurons were modeled with three various training epochs of 100, 200, and 300 to generate several combinations of LSTM prediction networks. Moreover, the training optimizer used is "sgdm" or the Stochastic Gradient Descent with momentum with a set initial learning rate of 0.001, a mini-batch size of 128 to learn the common patterns as important features, and a gradient threshold of 1.

On the other hand, Extreme Learning Machine or ELM is also applied to generate prediction models. ELMs are feedforward neural networks having one or more layers of hidden nodes that are used to analyze data and predict values [41]. These hidden nodes' parameters require no adjustment for selecting features, compression, clustering, classification, or sparse estimation [41]. To ensure lower error rates, weights to these concealed nodes may be assigned using the stochastic projection method, or these nodes can be passed down from their predecessors and not changed. In contrast to conventional gradient-based methods of learning for feedforward neural networks, ELM offers intriguing and important characteristics [42]. In comparison to

Hyperparameters	Value
Number of Hidden Neurons in Layer 1	500, 1000, 1500
Number of Hidden Neurons in Layer2	700, 500, 300
Number of Hidden Neurons in Layer 3	300, 200, 100
Number of Epochs	100, 200, 300
Training Optimizer	"sgdm"
Initial Learning Rate	0.001
Mini-Batch size	128
Gradient Threshold	1

Table 1.
Hyperparameters used in LSTM prediction modeling of latitude and longitude.

Hyperparameters	Value
Number of Hidden Neurons	100, 200, 300, 400, 500, 600, 700, 800, 900, 1000
Activation Function	"radbas"

Table 2.
Hyperparameters used in ELM prediction modeling of latitude and longitude.

gradient-based learning, ELM learning progresses far more quickly and performs well in generalization [43]. The hyperparameters used in the simulation of ELM models for latitude and longitude prediction are summarized in **Table 2**. A single-layer ELM is used, thus producing various models by simulating the different numbers of hidden neurons, which arc given as 100, 200, 300, 400, 500, 600, 700, 800, 900, and 1000 while the selected activation function applied is the "radbas" or the radial basis function for good generalization and fast training.

2.5 Evaluation metrics for prediction model performance

The performance of the developed prediction models for latitude and longitude correction was evaluated using mean square error (MSE), root mean square error (RMSE), coefficient of determination (R^2), and mean absolute error (MAE).

MSE calculates the average difference of squares between predicted and true values. MSE is used to assess the model's quality based on predictions made across the entire training dataset versus the true label/output value. Lower MSE values suggest that the model is more accurate, and this is defined mathematically by (1):

$$MSE = \frac{1}{n}\sum_{i=1}^{n}\left(\hat{y}_i - y_i\right)^2 \tag{1}$$

The average difference between expected and actual values in a dataset is measured using the root mean square error (RMSE). The RMSE measures how distributed the residuals are, showing how closely the observed data clusters around the predicted values. Mathematically, RMSE is calculated as the square root of the MSE.

The coefficient of determination (R^2) is a measure of how well the values fit in comparison to the original values. It calculates the percentage of the total variation in the variable that is dependent which can be explained by the model's independent variables. The value is obtained as a percentage and ranges from 0 to 1. The greater the value, the better the model. It is calculated using (2).

$$R^2 = 1 - \frac{\sum_{i=1}^{n}\left(y_i - \hat{y}_i\right)^2}{\sum_{i=1}^{n}\left(y_i - \bar{y}_i\right)^2} \tag{2}$$

Finally, MAE indicates the difference between the true and predicted values, which is calculated by averaging the absolute difference over a given data set. It is typically utilized when measuring performance using continuous variable data.

It returns a linear number that equalizes the weighted individual differences. The smaller the value, the greater the performance of the model. It is computed using (3).

$$MAE = \frac{\sum_{i=1}^{n} \left| y_i - \hat{y}_i \right|}{n} \tag{3}$$

where n is equal to the number of data points, y_i is the observed values, and $\hat{y}_i$ is the predicted values.

3. Results and discussion

3.1 Analysis of results of ANFIS-based GPS-IMU data correction models

Different combinations of ANFIS optimization algorithms and several epochs are applied to train and test the data to select the best ANFIS model for GPS latitude and longitude correction. The 2521 input and output data rows were divided into training, validation, and test data. About 55% is used as training data, 25% for validation, and 20% for testing. The input data used are collected actual data, specifically the GPS latitude and longitude coordinates, 3-axis coordinates of the accelerometer, and 3-axis coordinates of the magnetometer, while the output data used is the extracted reference GPS latitude and longitude from Google Maps. The simulated models are further quantified numerically using training RMSE, validation RMSE, and testing RMSE. Presented in **Table 3** is the summary of simulated ANFIS models with different hyperparameters for latitude correction. Model 1 corresponds to the ANFIS latitude correction model, which resulted in four models known as models 1A, 1B, 1C, and 1D. In model 1A, hybrid optimization is applied, and after 100 epochs, the model training has stopped attaining training RMSE of 0.009121, validation RMSE of 0.008567, and testing RMSE of 0.012093. On the other hand, model 1C with the backpropagation algorithm applied was trained for 100 epochs, achieving training RMSE of 1.101670, validation MSE of 1.136290, and testing RMSE of 4.383700. Similarly, model 1D with the backpropagation algorithm at 300 epochs obtained a low training RMSE of 1.090430, validation RMSE of 1.114220, and testing RMSE of 4.473100. However, model 1B with a hybrid optimization algorithm achieved the lowest training RMSE of 0.008770, 0.008300 validation RMSE, and 0.011814 testing RMSE at 300 training epochs. It is evident that model 1B significantly exhibited the best training and test results out of the other models for GPS latitude correction using the simulated ANFIS tool.

Model 1	Optimization Algorithm	Epochs	Training RMSE	Validation RMSE	Testing RMSE
1A	Hybrid	100	0.009121	0.008567	0.012093
1B	Hybrid	300	0.008770	0.008300	0.011814
1C	Backpropagation	100	1.101670	1.136290	4.383700
1D	Backpropagation	300	1.090430	1.114220	4.473100

Table 3.
Summary of simulated ANFIS models with different hyperparameters for latitude correction.

Model 2	Optimization Algorithm	Epochs	Training RMSE	Validation RMSE	Testing RMSE
2A	Hybrid	100	0.007361	0.007100	0.015321
2B	Hybrid	300	0.007361	0.007100	0.015321
2C	Backpropagation	100	0.766548	0.773640	3.360600
2D	Backpropagation	300	0.915090	0.905851	2.565400

Table 4.
Summary of simulated ANFIS models with different hyperparameters for longitude correction.

Moreover, the summary of simulated ANFIS models for different combinations of hyperparameters for GPS longitude correction is presented in **Table 4**. In contrast to model 1, model 2 represents the ANFIS longitude correction model, which resulted in models 2A, 2B, 2C, and 2D models. Model 2C with the backpropagation algorithm applied was trained for 100 epochs, achieving a high training RMSE of 0.766548, validation MSE of 0.773640, and testing RMSE of 3.360600. But model 2D with a backpropagation algorithm at 300 epochs obtained the highest training RMSE of 0.915090, validation RMSE of 0.905851, and testing RMSE of 2.565400. With a hybrid optimization algorithm applied, model 1A and model 2B resulted in the lowest and same values of training RMSE, validation RMSE, and testing RMSE of 0.007361, 0.007100, and 0.015321, respectively. However, based on the training results, model 2B seems to overfit the training data and has a more complex model with higher computational resources required for training because of the higher training epoch value. Thus, the superior performance of model 2A in correcting GPS longitude is apparent as it has shown significantly better training and test results than the other models.

3.2 Analysis of results of GPS-IMU data correction models using LSTM and ELM

To compare and validate the results of the simulated ANFIS models, different LSTM and ELM models were also simulated to achieve their best-offered models. The 2521 input and output data rows were split into training, validation, and test data, as with ANFIS models. About 55% of the data is used for training, 25% for validation, and 20% for testing. The input data consists of actual data, particularly GPS latitude and longitude coordinates, accelerometer 3-axis coordinates, and magnetometer 3-axis coordinates. In contrast, the output data is the extracted reference GPS latitude and longitude from Google Maps. The simulations of LSTM models for latitude and longitude prediction were done in MATLAB. A total of nine combinations of LSTM networks using the different set hyperparameters have been modeled. From the different combinations, the results of the best combinations of the LSTM model are shown in **Table 5**. The table shows that an LSTM network comprised of 1500–700-300 hidden neurons for each of the three hidden layers with 300 epochs is the best

Parameter	LSTM Network	Epochs	Training RMSE	Validation RMSE	Testing RMSE
Latitude	1500–700-300	300	0.106777	0.100347	0.101889
Longitude	1500–700-300	200	0.145039	0.149588	0.149149

Table 5.
Results of best LSTM models for latitude and longitude correction.

Parameter	ELM Hidden Neuron	Training RMSE	Validation RMSE	Testing RMSE
Latitude	100	0.072205	0.078897	0.079802
Longitude	100	0.109437	0.119175	0.138129

Table 6.
Results of best ELM models for latitude and longitude correction.

model for latitude correction among the other simulated models. This model obtained the least training RMSE of 0.106777, validation RMSE of 0.100347, and testing RMSE of 0.101889. Moreover, the best simulated LSTM model for longitude correction is the combination of 1500–700-300 hidden neurons for each of the three hidden layers with less training epochs of 200. Thus, this selected model obtained the least training RMSE of 0.145039, validation RMSE of 0.149588, and testing RMSE of 0.149149.

Compared to the LSTM models, **Table 6** also provides the results of the highest-performing simulated ELM models for latitude and longitude correction. From the results, the ELM network with 100 hidden neurons indicates the least training, validation, and testing RMSE for latitude and longitude correction. The training RMSE of the selected highest-performing ELM model for latitude correction is 0.72205, the validation RMSE is 0.78897, and its testing RMSE is 0.79802. On the other hand, the training RMSE of the highest-performing ELM model for longitude prediction is 0.109437, with a validation RMSE of 0.119175 and training RMSE of 0.138129.

3.3 Comparison of results between ANFIS, LSTM, and ELM

Using 25% of the collected data as validation data for model evaluation, the selected best ANFIS model performance is compared to the selected highest-performing LSTM and ELM models for latitude correction which is presented in **Table 7**. In terms of MSE, the selected best ANFIS model, model 1B from **Table 3**, showed significantly superior results compared to the LSTM and ELM models, which is 0.000069. Other evaluation metrics such as the R^2 and MAE of 0.995479 and 0.000375, respectively, also signify that the best ANFIS model still offers superior performance.

Furthermore, among the simulated highest-performing ANFIS, LSTM, and ELM prediction models for longitude correction in **Table 8**, the ANFIS model, which is model 2A from **Table 4**, still has the most superior performance with attained MSE, R^2, and MAE of 0.000050, 0.997675, and 0.000042, respectively.

A scatter plot is presented to visualize the performance of the ANFIS models using the validation data (**Figures 11** and **12**) to compare further the relationship between the reference latitude/longitude and the predicted corrected latitude/longitude. The given plot is clear and concise, indicating a strong relationship between the predicted and response variables. The predicted latitude/longitude values are close enough to the reference latitude/longitude values as the points cluster around the trend line.

Model	MSE	R^2	MAE
ANFIS	0.000069	0.995479	0.000375
LSTM	0.010070	0.250617	0.006071
ELM	0.001746	0.931812	0.002186

Table 7.
Summary of the evaluation metrics for the selected best models for GPS latitude correction.

Model	MSE	R^2	MAE
ANFIS	0.00005	0.997675	0.000042
LSTM	0.022376	0.110682	0.001096
ELM	0.004519	0.855484	0.000425

Table 8.
Summary of the evaluation metrics for the selected best models for GPS longitude correction.

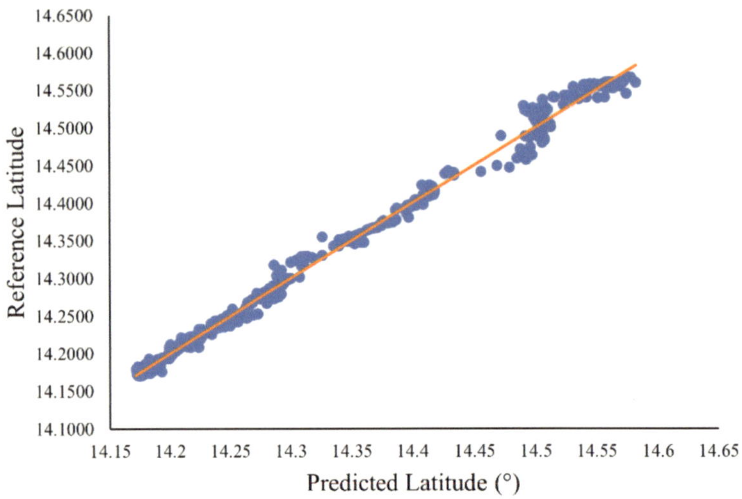

Figure 11.
The resulting scatter plot of the predicted corrected latitude of the selected most superior ANFIS model versus the reference latitude data.

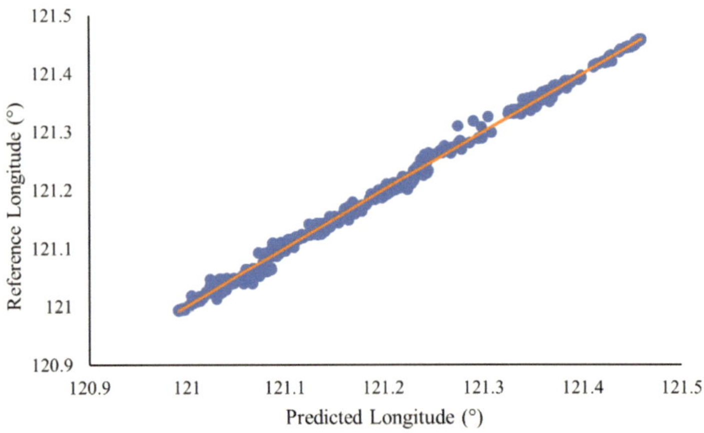

Figure 12.
The resulting scatter plot of the predicted corrected longitude of the selected most superior ANFIS model versus the reference longitude data.

Therefore, the simulated ANFIS models still outperformed the LSTM and ELM models, which proved that these models can combine the advantageous features of neural networks and fuzzy logic in one framework. In giving better accuracy in predicting the corrected latitude and longitude. The comparison of results also proved that ANFIS is a promising method for localization and tracking vehicles utilizing GPS and IMU data. In terms of generalization, ANFIS has demonstrated high generalization capability, which increases its robustness and accuracy when transforming fuzzy sets into crisp inputs [22].

3.4 Visualization results of the actual collected dataset by applying the selected best ANFIS models

To confirm the robustness of the selected best models, 2521 collected actual GPS-IMU datasets obtained from the conducted testing from Lumban, Laguna to DLSU, without the need of splitting the data, were used to test and visualize the ANFIS model performances. These actual datasets consist of the uncorrected GPS latitude and longitude coordinates. On the other hand, the corrected GPS latitude and longitude were predicted using the selected highest-performing ANFIS models (ANFIS model 1B for latitude and model 2B for longitude). The visualized map of the plotted actual GPS latitude and longitude dataset and the predicted corrected GPS latitude and longitude through ANFIS are presented in **Figure 13**. It shows that the uncorrected set of GPS latitude and longitude was too outlying and very distant from the predicted corrected output values generated by the ANFIS models. This signifies that the selected ANFIS models have the best performances to predict the correct GPS latitude and longitude.

The response of the predicted corrected GPS coordinates using the selected ANFIS models concerning the reference route and collected actual GPS coordinates are shown (**Figure 14**). The result of the plotted corrected GPS latitude and longitude coordinates (in green dotted line) is almost near the reference GPS coordinates (in blue line) compared to the collected actual GPS latitude and longitude (in red pin markers). According to the results, the proposed ANFIS models, that is, the fuzzy system combined with neural network capabilities, achieved better error reduction without the need to identify the system's noise type, as it was trained on the

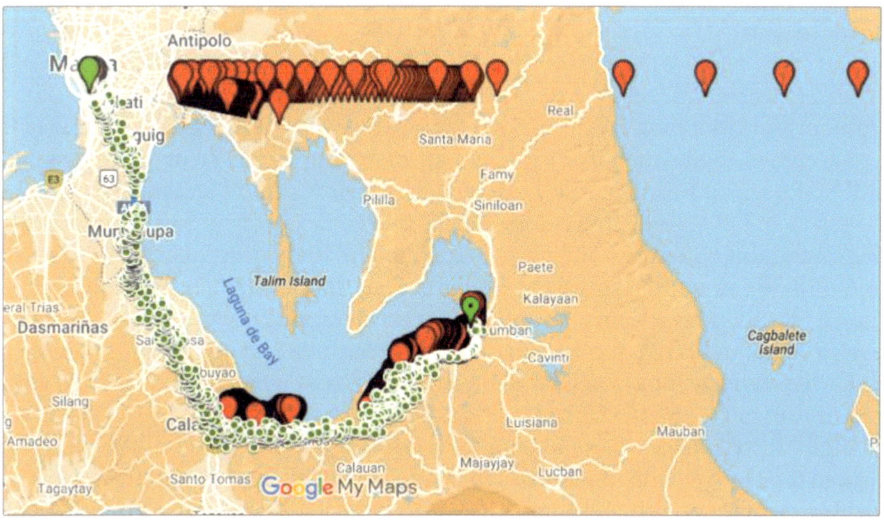

Figure 13.
Visualization maps of the uncorrected GPS coordinates (red pin marker) versus the predicted corrected GPS coordinates (green dots).

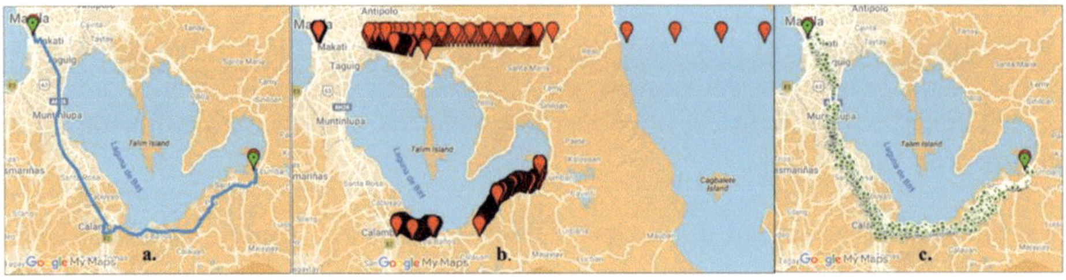

Figure 14.
Resulting visualization maps for comparison of the (a) reference GPS coordinates, (b) uncorrected GPS coordinates, and (c) corrected GPS coordinates using the ANFIS models.

data region. This makes it a more convenient and cost-effective option than other state-of-the-art approaches. In the given noisy collected actual dataset, the selected ANFIS models have been proven to address the challenging and nonlinear problems while minimizing complexity in computation [27]. Because ANFIS models may be implemented in real-time, they are well suited for applications that demand rapid and precise data processing, particularly suitable for the localization and tracking of vehicle position [27]. Integrating the ANFIS models into the capacitive resistivity underground imaging towed antenna system can offer great significance in mapping the precise location of the surveyed underground utility objects.

However, verifying further the ANFIS models' performance requires the recollection of new GPS latitude and longitude datasets and testing this new data to the simulated ANFIS models. This is considered as the lacking approach of this study that can be done for future research. Additionally, to maximize the ANFIS model performance, it suggests retraining and adjusting the hyperparameters when tested in newly collected actual data. Also, the study's results can be further evaluated by comparing them with other methods of sensor fusion such as the use of other machine learning models which will be the next direction of the paper.

4. Conclusion

The CR method utilizes GPS to create accurate maps quickly and with less equipment and labor than traditional surveying. However, errors can occur due to GPS sensor accuracy, digital map quality, and map-matching errors. To improve accuracy, an IMU-GPS sensor fusion method can be used. Environmental factors can still cause GPS sensors to fail, so reducing errors in GPS receiver accuracy is crucial for correct underground utility location and map matching. The study proposes using fuzzy logic with ANFIS to correct the latitude and longitude of the CR vehicle's position by integrating RTK-GPS sensor data with IMU linear acceleration and magnetometer data. The ANFIS tool trains two fuzzy systems, one for latitude and one for longitude correction. The developed combined GPS-IMU circuit was tested by conducting field testing from Lumban to De La Salle University Manila Campus to collect actual GPS latitude and longitude, triaxial accelerometer, and triaxial magnetometer data. The study evaluated different ANFIS models with varying hyperparameters, and the selected model for latitude correction is model 1B with a hybrid optimization algorithm at 300 training epochs. This model achieved the lowest training RMSE of 0.008770, validation RMSE of 0.008300, and testing RMSE of 0.011814 at 300 training epochs. Model 1B showed the lowest MSE of 0.000069, highest R^2 of 0.995479, and lowest MAE of

0.000375 compared to other models, proving superior results. For ANFIS longitude correction, model 2B was selected, and a hybrid optimization algorithm was applied at 100 epochs, which resulted in the lowest training RMSE of 0.007361, 0.007100 validation RMSE, and 0.01532 testing RMSE. Among the four prediction models of model 2, model 2A achieved the lowest MSE of 0.000050, the highest R^2 of 0.997675, and the lowest MAE of 0.000042 for longitude correction, demonstrating the best results.

The selected best ANFIS models' performances were then validated by comparing them to simulated LSTM and ELM models; however, the ANFIS models still outperformed the two other models. The ANFIS models were also tested on the collected actual dataset for verification of results. The visualized map obtained from this simulation test revealed that the uncorrected GPS data points were significantly distant from the target GPS reference values compared to the ANFIS corrected output. This indicates that the ANFIS models proved to combine the benefits of neural networks with fuzzy logic in a single structure for predicting corrected latitude and longitude with greater accuracy. The comparison of findings also demonstrated that ANFIS is a potential solution for vehicle localization and tracking using GPS and IMU data, making it suitable to be integrated into capacitive resistivity underground imaging system and can be extremely useful in mapping the precise location of the investigated subterranean utility objects.

To further validate the ANFIS models' performance, additional GPS latitude and longitude datasets must be collected and tested against the simulated ANFIS models. This is considered the study's lacking strategy that can be done for future research. Furthermore, it suggests retraining and adjusting the hyperparameters when tested on newly obtained actual data to maximize the ANFIS model performance. Lastly, the study's results can be further examined by comparing them to other ways of sensor fusion, such as the employment of other machine learning models, which will be the paper's next goal.

Acknowledgements

The authors would like to thank the Department of Science and Technology – Philippine Council for Industry, Energy and Emerging Technology Research and Development (DOST-PCIEERD) and the Intelligent Systems Laboratory of the De La Salle University for all the granted support.

Author details

Elmer Dadios[1,2*], Jonah Jahara Baun[2,3], Mike Louie Enriquez[1,2], Adrian Genevie Janairo[2,3], Ronnie Concepcion II[1,2], Joseph Aristotle De Leon[1,2], Kate Francisco[1,2], Andres Philip Mayol[2,4], Argel Bandala[2,3] and Ryan Rhay Vicerra[1,2]

1 Department of Manufacturing Engineering and Management, De La Salle University, Manila, Philippines

2 Center for Engineering and Sustainability Development Research, De La Salle University, Manila, Philippines

3 Department of Electronics and Computer Engineering, De La Salle University, Manila, Philippines

4 Department of Mechanical Engineering, De La Salle University, Manila, Philippines

*Address all correspondence to: elmer.dadios@dlsu.edu.ph

References

[1] Ducut JD, Alipio M, Go PJ, Concepcion R II, Vicerra RR, Bandala A, et al. A review of electrical resistivity tomography applications in underground imaging and object detection. Displays. 2022;**73**:85-94

[2] Suherman E, Wijayanto AH, Rambe N, Mubarakah YS, Al-Akaidi M. Underground object detection based on radio propagation characteristics. Journal of Theoretical and Applied Information Technology. 2021;**99**(1):63-71

[3] Kuras O, Beamish D, Meldrum PI, Ogilvy RD. Fundamentals of the capacitive resistivity technique. Geophysics. 2006;**71**(3):29-41

[4] Janairo AG, Baun JJ, Concepcion R, Relano R, Francisco K, Enriquez ML, et al. Optimization of subsurface imaging antenna capacitance through geometry modeling using Archimedes, Lichtenberg and Henry gas solubility metaheuristics. In: 2022 IEEE International IOT, Electronics and Mechatronics Conference (IEMTRONICS). Toronto, ON, Canada: Institute of Electrical and Electronics Engineers Inc.; 2022. pp. 1-8

[5] Francisco K et al. Systematic analysis and proposed AI-based technique for attenuating inductive and capacitive parasitics in low and very low frequency antennas. In: IEEE International IOT, Electronics and Mechatronics Conference. Toronto, ON, Canada: Institute of Electrical and Electronics Engineers Inc.; 2022

[6] Francisco K et al. Analytical hierarchical process-based material selection for trailer body frame of an underground imaging system. In: 2021 IEEE 13th International Conference on Humanoid, Nanotechnology, Information Technology, Communication and Control, Environment, and Management. Manila, Philippines: Institute of Electrical and Electronics Engineers Inc.; 2021

[7] Baun JJ et al. Hybrid stochastic genetic evolution-based prediction model of received input voltage for underground imaging applications. In: 2023 8th International Conference on Business and Industrial Research (ICBIR). Bangkok, Thailand: IEEE; 2023. pp. 549-555. DOI: 10.1109/ICBIR57571.2023.10147464

[8] Lo Monte L, Erricolo D, Soldovieri F, Wicks MC. Underground imaging of irregular terrains using RF tomography. In: 2009 3rd IEEE International Workshop on Computational Advances in Multi-Sensor Adaptive Processing (CAMSAP). Aruba, Netherland Antille: IEEE; 2009. pp. 229-232

[9] Nandakumar N, Alex TK, Ajayakumar K, Rao KS. Capacitive resistivity imaging: Overview, modeling and applications. Journal of Selected Topics in Applied Earth Observations and Remote Sensing (IEEE). Sep 2010;**3**(3):376-390

[10] Hsu PC, Nurrachman MA. GPS limitations and solutions. In: 2019 IEEE International Conference on Systems, Man and Cybernetics (SMC). Bari, Italy: IEEE; 2019. pp. 2355-2360

[11] Li J, Li Y, Li B, Yang D. Performance analysis and improvement of GPS positioning accuracy. IEEE Access. 2021;**9**:36368-36380

[12] Islam MS, Rahman MH, Rashid MA, Khan MFR. Challenges and limitations of GPS: A review. In: 2017 International

Conference on Electrical, Computer and Communication Engineering (ECCE). Cox's Bazar, Bangladesh: IEEE; 2017. pp. 548-552. DOI: 10.1109/ECCE.2017.8307247

[13] Panda SS, Padhy PK. Map-matching algorithms for vehicle navigation: A survey. Journal of Intelligent Transportation Systems. 2014;**18**(2):99-117. DOI: 10.1080/15472450.2013.841344

[14] Zhou Y, Li H, Ding L. IMU/GPS fusion with extended Kalman filter for autonomous vehicle navigation. In: 2017 IEEE International Conference on Mechatronics and Automation (ICMA). Takamatsu: IEEE; 2017. pp. 648-653. DOI: 10.1109/ICMA.2017.8015795

[15] Shien C, Huang C, Liu J, Chen X, Weijinya U, Shi G. Integrated navigation accuracy improvement algorithm based on multi-sensor fusion. In: 2019 IEEE The 2nd International Conference on Micro/Nano Sensors for AI, Healthcare, and Robotics (NSENS). Shenzhen, China: IEEE; 2019. pp. 54-57. DOI: 10.1109/NSENS49395.2019.9293950

[16] Farhad M, Mosavi M, Abedi A, Mohammadi K. Increasing the resistance of GPS receivers by using a fuzzy smart estimator in weak signal conditions. The Journal of Navigation. 2020;**73**(5):991-1013. DOI: 10.1017/S0373463320000132

[17] Correa-Caicedo PJ, Barranco-Gutiérrez AI, Guerra-Hernandez EI, Batres-Mendoza P, Padilla-Medina JA, Rostro-González H. An FPGA-based architecture for a latitude and longitude correction in autonomous navigation tasks. Measurement. 2021;**182**:109757. DOI: 10.1016/j.measurement.2021.109757

[18] Kathpalia N, Gulati T. Enhancement of GPS accuracy using combination of active antenna ground plane enhancement, sensor fusion compressing 3 axis gyro accelerometer & artificial intelligence. Mathematical Statistician and Engineering Applications. 2022;**71**(3s2):2067-2083

[19] Tehrani M, Nariman-Zadeh N, Masoumnezhad M. Adaptive fuzzy hybrid unscented/H-infinity filter for state estimation of nonlinear dynamics problems. Transactions of the Institute of Measurement and Control. 2019;**41**(6):1676-1685

[20] Woo R, Yang E-J, Seo D-W. A fuzzy-innovation-based adaptive Kalman filter for enhanced vehicle positioning in dense urban environments. Sensors. 2019;**19**(5):1142. DOI: 10.3390/s19051142

[21] Zhu J, Tang Y, Shao X, Xie Y. Multisensor fusion using fuzzy inference system for a visual-IMU-wheel odometry. IEEE Transactions on Instrumentation and Measurement. 2021;**70**:2505216. DOI: 10.1109/TIM.2021.3105867

[22] Chopra S, Dhiman G, Sharma A, Shabaz M, Shukla P, Arora M. Taxonomy of adaptive neuro-fuzzy inference system in modern engineering sciences. Computational Intelligence and Neuroscience. 2021;**2021**:6455592. DOI: 10.1155/2021/6455592

[23] Sun R, Hsu L, Xue D, Zhang G, Ochieng W. GPS signal reception classification using adaptive neuro-fuzzy inference system. The Journal of Navigation. 2019;**72**(3):685-701. DOI: 10.1017/S0373463318000899

[24] Correa Caicedo P, Rostro-Gonzalez H, Rodriguez-Licea M, Gutiérrez-Frías Ó, Herrera-Ramírez C, Méndez-Gurrola I, et al. GPS data correction based on fuzzy logic for tracking land vehicles. Mathematics. 2021;**9**(21):2818. DOI: 10.3390/math9212818

[25] Duan Y, Li H, Wu S, Zhang K. INS error estimation based on an ANFIS and

its application in complex and covert surroundings. ISPRS International Journal of Geo-Information. 2021;**10**(6):399. DOI: 10.3390/ijgi10060388

[26] Ajani OS, El-Hussieny H. An ANFIS-based human activity recognition using IMU sensor fusion. In: 2019 Novel Intelligent and Leading Emerging Sciences Conference (NILES). Giza, Egypt: Institute of Electrical and Electronics Engineers Inc.; 2019. pp. 34-37. DOI: 10.1109/NILES.2019.8909289

[27] El Shafie A, Hussain A, Eldin AEN. ANFIS-based model for real-time INS/GPS data fusion for vehicular navigation system. In: 2009 International Conference on Computer Technology and Development. Kota Kinabalu, Malaysia: Institute of Electrical and Electronics Engineers Inc.; 2009. pp. 278-282. DOI: 10.1109/ICCTD.2009.42

[28] Hosseinyalamdary S. Deep Kalman filter: Simultaneous multi-sensor integration and modelling, a GNSS/IMU case study. Sensors (Basel). 2018;**18**(5):1316. DOI: 10.3390/s18051316

[29] De Leon JA et al. Robotic simulation and GPS position estimation of a towed CRI equipment using Savitzky-Golay filter. In: IEEE Region 10 Annual International Conference, Proceedings/TENCON. Hong Kong, Hong Kong: Institute of Electrical and Electronics Engineers Inc.; 2022. DOI: 10.1109/TENCON55691.2022.9977685

[30] Stangebye T, Mohr T, Vallenti A, Grauff M, Koziol S. Custom real-time-kinematics positioning system testbed for mobile robot localization. In: Proc. of the IEEE Dallas Circuits and Systems Conference. Dallas, TX, USA: Institute of Electrical and Electronics Engineers Inc.; 2020. pp. 221-224. DOI: 10.1109/DCAS51144.2020.9330659

[31] Broekman A, Gräbe J. A low-cost, mobile real-time kinematic geolocation service for engineering and research applications specifications table. HardwareX. 2021;**10**:e00203. DOI: 10.17605/OSF.IO/QH3V7

[32] Abdelfatah R, Moawad A, Alshaer N, Ismail T. UAV tracking system using integrated sensor fusion with RTK-GPS. In: 2021 International Mobile, Intelligent, and Ubiquitous Computing Conference, MIUCC 2021. Cairo, Egypt: Institute of Electrical and Electronics Engineers Inc.; 2021. pp. 352-356. DOI: 10.1109/MIUCC52538.2021.9447646

[33] Brossard M, Barrau A, Bonnabel S. AI-IMU dead-reckoning. IEEE Transactions on Intelligent Vehicles. 2020;**5**(4):585-595. DOI: 10.1109/TIV.2020.2980758

[34] Fazeli H, Samadzadegan F, Dadrasjavan F. Evaluating the potential of RTK-UAV for automatic point cloud generation in 3D rapid mapping. In: International Archives of the Photogrammetry, Remote Sensing and Spatial Information Sciences - ISPRS Archives. Prague, Czech Republic: International Society for Photogrammetry and Remote Sensing; 2016. pp. 221-226. DOI: 10.5194/isprsarchives-XLI-B6-221-2016

[35] Lauguico S, Baldovino R, Concepcion R, Alejandrino J, Tobias RR, Dadios E. Adaptive neuro-fuzzy inference system on aquaphotomics development for aquaponic water nutrient assessments and analyses. In: ICITEE 2020 - Proceedings of the 12th International Conference on Information Technology and Electrical Engineering. Yogyakarta, Indonesia: Institute of Electrical and Electronics Engineers Inc.; 2020. pp. 317-322. DOI: 10.1109/ICITEE49829.2020.9271736

[36] Jang JSR. ANFIS: Adaptive-network-based fuzzy inference system. IEEE Transactions on Systems, Man and Cybernetics. 1993;**23**(3):665-685. DOI: 10.1109/21.256541

[37] Francisco KG, Relano RS, Concepcion RS, Enriquez ML, Vicerra RR, Baun JJ, et al. Adaptive neuro-fuzzy active spring control of a vibration isolator for towed capacitive resistivity imaging single trailer. In: 2022 IEEE 14th International Conference on Humanoid, Nanotechnology, Information Technology, Communication and Control, Environment, and Management (HNICEM). Boracay Island, Philippines: Institute of Electrical and Electronics Engineers Inc.; 2022. pp. 1-6. DOI: 10.1109/HNICEM57413.2022.10109420

[38] Panoiu M, Panoiu C, Lihaciu IL. Adaptive neuro fuzzy system for modelling and prediction of distance pantograph catenary in railway transportation. IOP Conference Series: Materials Science and Engineering. 2018;**294**(1):1-7. DOI: 10.1088/1757-899X/294/1/012073

[39] Arafeh L, Singh H, Putatunda SK. A neuro fuzzy logic approach to material processing. IEEE Transactions on Systems, Man, and Cybernetics Part C (Applications and Reviews). 1999;**29**(3):362-370. DOI: 10.1109/5326.777072

[40] Prakash S, Jalal AS, Pathak P. Forecasting COVID-19 pandemic using prophet, LSTM, hybrid GRU-LSTM, CNN-LSTM, Bi-LSTM and stacked-LSTM for India. In: 2023 6th International Conference on Information Systems and Computer Networks (ISCON). Mathura, India: Institute of Electrical and Electronics Engineers Inc.; 2023. pp. 1-6. DOI: 10.1109/ISCON57294.2023.10112065

[41] Vashisht S, Kumar P, Trivedi MC. Improvised extreme learning machine for crop yield prediction. In: 2022 3rd International Conference on Intelligent Engineering and Management (ICIEM). London, United Kingdom: Institute of Electrical and Electronics Engineers Inc.; 2022. pp. 754-757. DOI: 10.1109/ICIEM54221.2022.9853054

[42] Sa'ad Kusrini MI, Mustafa MS. Student prediction of drop out using extreme learning machine (ELM) algorithm. In: 2020 2nd International Conference on Cybernetics and Intelligent System (ICORIS). Manado, Indonesia: Institute of Electrical and Electronics Engineers Inc.; 2020. pp. 1-6. DOI: 10.1109/ICORIS50180.2020.9320831

[43] Zhu QY, Qin AK, Suganthan PN, Huang GB. Evalutionary extreme learning machine. Pattern Recognition. 2005;**38**(10):1759-1763. DOI: 10.1016/j.patcog.2005.03.028